RECORDS AND BRANCHING PROCESSES

RECORDS AND BRANCHING PROCESSES

MOHAMMAD AHSANULLAH
AND
GEORGE P. YANEV
EDITORS

Nova Science Publishers, Inc.
New York

For permission to use material from this book please contact us:
Telephone 631-231-7269; Fax 631-231-8175
Web Site: http://www.novapublishers.com

Library of Congress Cataloging-in-Publication Data

Records and branching processes / Mohammad Ahsanullah and George P. Yanev (editors).
p. cm.
ISBN 978-1-60456-526-3 (hardcover)
1. Distribution (Probability theory) 2. Branching processes. 3. Mathematical statistics. I. Ahsanullah, M. (Mohammad) II. Yanev, George P.
QA273.6.R43 2008
519.2'4--dc22 2008012148

Published by Nova Science Publishers, Inc. ✢ New York

CONTENTS

PREFACE

Record values and branching processes are active and rich research areas of applied probability and statistics. Over the last few decades, numerous interesting articles on these topics appeared, dealing with theoretical problems as well as a number of new applications. It is both an honor and professional interest for the editors to coordinate publishing works contributed by eminent researchers from all over the world. This book presents a selection of presentations of new developments and survey papers on the subjects of record values and branching processes. It is written at an intermediate level which requires knowledge of probability theory and mathematical statistics.

The book contains seven chapters. The first chapter describes some basic properties of record values and inter-record times. Representations of records as functions of independent random variables as well as some characterizations of distributions are presented in this chapter. The second chapter discusses characterizations of distributions based on differences of records. Chapter 3 presents a review of results on branching processes with regeneration at zero. Recent results illustrate a new probabilistic technique for obtaining limit theorems. A generalization of the Foster-Pakes process with time non-homogeneous state-dependent immigration is also given. Chapter 4 gives a unified treatment of conditional limit theorems for the non-supercritical continuous time and state branching process. A stationary measure is shown to exist if and only if extinction is possible. Chapter 5 presents recent results on exceedances and genealogy of branching processes. In this chapter various limit theorems for the index of the first process in a sequence of branching processes which exceeds a threshold are presented. Chapter 6 reviews findings on extremes related to the family size in branching processes. Some of the results are illustrated with new examples. Chapter 7 is a comprehensive survey of non-parametric statistical results for discrete-time single-type branching processes under different observational schemes. Special attention is given to processes with increasing number of ancestors. This construction leads to a unified estimation theory for subcritical, critical, and supercritical processes under different sample schemes.

Summer research grant and sabbatical leave from Rider University enabled the first author to complete a part of the book. He also thanks his wife Masuda for her support and encouragement during the preparation of the book. The second author thanks his wife Reneta for her constant support.

We are grateful to Nova Publishers for willingness to publish the manuscript.

Mohammad Ahsanullah
George Yanev
November 2007

CONTRIBUTORS AND AFFILIATION

Mohammad Ahsanullah
Department of Management Sciences, Rider University, 2083 Lawrenceville Road, Lawrenceville, NJ 08648, USA, ahsan@rider.edu.

Ilham Akhundov
Statistics and Actuarial Science,
University of Waterloo 200 University Avenue West Waterloo, Ontario, Canada N2L 3G1, iakhundo@math.uwaterloo.ca.

Kosto V. Mitov
Aviation Faculty, National Military University, 5856 Dolna Mitropolia, Bulgaria, kmitov@af-acad.bg.

Valery B. Nevzorov
Department of Mathematics and Mechanics, Saint Petersburg State University, Starii Petergoff, Universitetskii Prospekt 28, Saint Petersburg, 198504 Russia, decanat@math.spbu.ru.

Anthony G. Pakes
School of Mathematics \& Statistics, University of Western Australia, 35 Stirling Highway, Crawley, WA, Australia 6009, pakes@maths.uwa.edu.au.

Ibrahim Rahimov
Department of Mathematical Sciences, King Fahd University of Petroleum and Minerals, P.O. Box 1339, Dhahran, 31261 Saudi Arabia and Romanovski Institute of Mathematics, Uzbek Academy of Sciences, 700143 Tashkent, Uzbekistan, rahimov@kfupm.edu.sa.

George P. Yanev
Department of Mathematics, University of Texas - Pan American, 1201 West University Drive, Edinburg, Texas 78539, USA, gyanev@utpa.edu.

Nikolay M. Yanev
Department of Probability and Statistics, Institute of Mathematics and Informatics, Bulgarian Academy of Sciences, 8 Acad. G. Bonchev Street, 1113 Sofia, Bulgaria, yanev@math.bas.bg.

In: Records and Branching Processes
Editor: M. Ahsanullah, G. P. Yanev, pp. 1-25
ISBN 978-1-60456-526-3

Chapter 1

RECORD STATISTICS

Mohammad Ahsanullah*
Department of Management Sciences, Rider University
2083 Lawrenceville Road, Lawrenceville, NJ 08648, USA

Abstract

In this chapter we consider record values in a sequence of independent and identically distributed random variables. Some basic properties of the record values are presented. Representations of records as functions of independent and identically distributed random variables are shown. Several characterization results of distributions using records are given.

Key words and phrases. Record Values, Weak Records, Inter Record Times, Representation of Records, Characterizations.

1 Definition of Record Values and Record Times

In this chapter record values of independent and identically distributed (i.i.d.) random variables are considered. Let $X_1, X_2, \dots$ be a sequence of i.i.d. absolutely continuous random variables with cumulative distribution function(cdf) $F(x)$ and probability density function (pdf) $f(x)$. Let Y_n=max(min) $\{X_1, X_2, \dots, X_n\}$, $n \geq 1$. We say that Y_j is an upper (lower) record value of $\{X_n,\ n \geq 1\}$, if $Y_j > (<) Y_{j-1},\ j \geq 1$. The indices at which the upper record values occur are given by record times $\{U(n),\ n > 0\}$, where $U(n) = \min\{j |\ j > U(n-1),\ X_j > X_{U(n-1)},\ n > 1\}$ and $U(1) = 1$. We will denote $L(n),\ n \geq 1$, as the indices where lower record values occur. By our assumption $U(1) = L(1) = 1$. By definition X_1 an upper as well as a lower record value. For i.i.d. discrete random variables we have two types of record values: ordinary and weak records. Let $X_1, X_2, \dots$ be a sequence of i.i.d. discrete random variables taking values on $0, 1, \dots$ with distribution function $F(x)$. The upper weak record times, $U_w(n)$, $n \geq 1$, are defined as follows: $U_w(1) = 1$ and $U_w(n+1) = \min\{j > U_w(n), X_j \quad \geq \max(X_1, \dots X_{j-1}\}$. The corresponding weak upper record values are defined as $X_{U_w}(n+1)$. If in the above expression, we replace $\geq$ by $>$, then we obtain ordinary upper record times and upper record values. We consider an example

*E-mail address: ahsan@rider.edu

to clarify the differences. Suppose the discrete random variables $X_1, X_2, \ldots$ have geometric distribution with probability mass function (pmf) of the following form for $k = 0, 1, 2, \ldots$,

$$p(k) = P(X_i = k) = pq^k, \quad 0 < p < 1, \quad q = 1 - p.$$

It can be shown that the marginal pmf of $X_{U(n)}$ is

$$P_n(x) = P(X_{U(n)} = x) = \binom{x}{n-1} p^n q^{x-n-1}, \quad x = n-1, n, \ldots; \quad n \geq 1.$$

The marginal pmf of $X_{U_w n}(x)$ is as follows

$$X_{U_w n}(x) = P(X_{U_w n} = x) = \binom{x+n-1}{n-1} p^n q^x, \qquad x = 0, 1, 2, \ldots; \quad n \geq 1.$$

In this chapter we consider continuous random variables unless mentioned otherwise.

2 Distribution of Record Values

Many properties of the record value sequence can be expressed in terms of the function $R(x) = -\ln \bar{F}(x)$, where $\bar{F}(x) = 1 - F(x)$. If we define $F_n(x)$ as the distribution function of $X_{U(n)}$ for $n \geq 1$, then $F_1(x) = P[X_{U(1)} \leq x] = F(x)$ and

$$\begin{aligned}
F_2(x) &= P[X_{U(2)} \leq x] \\
&= \int_{-\infty}^{x} \int_{-\infty}^{y} \sum_{i=1}^{\infty} (F(u))^{i-1} \, dF(u) \, dF(y) \\
&= \int_{-\infty}^{x} \int_{-\infty}^{y} \frac{dF(u)}{1 - F(u)} \, dF(y) \\
&= \int_{-\infty}^{x} R(y) \, dF(y)
\end{aligned}$$

If F(x) has a density f(x), then the pdf of $X_{U(2)}$ is $f_2(x) = R(x) f(x)$. The distribution function $F_3(x)$ of $X_{U(3)}$ is

$$\begin{aligned}
F_3(x) &= P(X_{U(3)} \leq x) \\
&= \int_{-\infty}^{x} \int_{-\infty}^{y} \sum_{i=0}^{\infty} (F(u))^{i} \, R(u) \, dF(u) \, dF(y) \\
&= \int_{-\infty}^{x} \int_{-\infty}^{y} \frac{R(u)}{1 - F(u)} \, dF(u) \, dF(y) \\
&= \int_{-\infty}^{x} \frac{(R(u))^2}{2!} \, dF(u).
\end{aligned}$$

The pdf $f_3(x)$ of $X_{U(3)}$ is $f_3(x) = (R(x))^2 f(x)/2!$, $-\infty < x < \infty$. It can similarly be shown that the cdf $F_n(x)$ of $X_{U(n)}$ is for $-\infty < x < \infty$

$$\begin{aligned}
F_n(x) &= \int_{-\infty}^{x} f(u_n) du_n \int_{-\infty}^{u_n} \frac{f(u_{n-1})}{1 - F(u_{n-1})} \, du_{n-1} \cdots \int_{-\infty}^{u_2} \frac{f(u_1)}{1 - F(u_1)} \, du_1 \\
&= \int_{-\infty}^{x} \frac{R^{n-1}(u)}{(n-1)!} \, dF(u).
\end{aligned}$$

This can be expressed as:

$$F_n(x) = \int_{-\infty}^{R(x)} \frac{u^{n-1}}{(n-1)!} e^{-u} du$$
$$= 1 - e^{-R(x)} \sum_{j=0}^{n-1} \frac{(R(x))^j}{j!}.$$

The pdf $f_n(x)$ of $X_{U(n)}$ is:

$$f_n(x) = \frac{(R(x))^{n-1}}{(n-1)!} f(x).$$

The joint pdf $f(x_1,x_2,\ldots,x_n)$ of the n record values $X_{U(1)}, X_{U(2)}, \ldots, X_{U(n)})$ is given for $-\infty < x_1 < x_2 < \cdots < x_{n-1} < x_n < \infty$, by:

$$f_{1,2,\ldots,n}(x_1,x_2,\ldots,x_n) = r(x_1)\, r(x_2)\ldots r(x_{n-1}) f(x_n)$$

where $r(x) = dR(x)/dx = f(x)/(1-F(x))$, $0 < F(x) < 1$. The function $r(x)$ is known as hazard rate.

The joint pdf of $X_{U(i)}$ and $X_{U(j)}$ is for $j > i$,

$$f_{i,j}(x_i,x_j) = \frac{(R(x_i))^{i-1}}{(i-1)!} r(x_i) \frac{(R(x_j)-R(x_i))^{j-i-1}}{(j-i-1)!} f(x_j)$$

for $-\infty < x_i < x_j < \infty$. In particular for $i = 1$ and $j = n$ we have

$$f_{1,n}(x_1,x_n) = r(x_1) \frac{(R(x_n)-R(x_1))^{n-2}}{(n-2)!} f(x_n)$$

for $-\infty < x_1 < x_n < \infty$.

Suppose we use the transformation $Y_1 = R(X_{U(i)})$ and $Y_2 = R(X_{U(i)})/R(X_{U(j)})$, $i < j$, then it can be shown that the pdf $f_2^*(y)$ of Y_2 is as follows:

$$f_2^*(y) = \frac{\Gamma(j)}{\Gamma(i)} \cdot \frac{1}{\Gamma(j-i)} \cdot y^{i-1} (1-y)^{j-i-1}, \qquad 0 < y < 1.$$

Thus Y_2 is distributed as Beta distribution with parameters i and j (i.e. B(i,j-i)). The mean and variance of Y_2 are $E(Y_2) = i/j$ and $Var(Y_2) = (ij)/[(j+1)j^2]$.

If we use the transformation $V_i = R(X_{U(i)})$, then the joint pdf of V_i, $i = 1,2,\ldots,n$, is $f(v_1,v_2,\cdots,v_n) = e^{-v_n}$, $0 < v_1 < v_2 < \cdots < v_n < \infty$. The joint distribution of V_m and V_r , $r > m$, is

$$f(v_m,v_r) = \frac{1}{\Gamma(m)} \cdot \frac{(v_r - v_m)^{r-m-1}}{\Gamma(r-m)} \cdot e^{-v_r}, \quad 0 < v_m < v_r < \infty.$$

and

$$E(V_k^l) = \int_0^{\infty} t^l \frac{1}{\Gamma(k)} t^{k-1} e^{-t} dt = \frac{\Gamma(k+l)}{\Gamma(k)}.$$

Thus $E(V_k) = Var(V_k) = k$. The conditional pdf of $X_{U(j)}| X_{U(i)} = x_i$ is for $j > i$,

$$f(x_j|X_{U(i)} = x_i) = \frac{f_{ij}(x_i, x_j)}{f_i(x_i)} = \frac{(R(x_j) - R(x_i))^{j-i-1}}{(j-i-1)!} \frac{f(x_j)}{1-F(x_i)}$$

for $-\infty < x_i < x_j < \infty$. In particular, for $j = i+1$ and $-\infty < x_i < x_{i+1} < \infty$, we have

$$f(x_{i+1}|X_{U(i)} = x_i) = \frac{f(x_{i+1})}{1-F(x_i)}$$

For $i > 0$, $1 \leq k < m$, the joint conditional pdf of $X_{U(i+k)}$ and $X_{U(i+m)}|X_{U(i)}$ is for $-\infty < z < x < y < \infty$

$$f_{i+k,,i+m}(x,y|X_{U(i)} = z)$$
$$= \frac{1}{\Gamma(m-k)} \frac{1}{\Gamma(k)} [R(y) - R(x)]^{m-k-1} [R(x) - R(z)]^{k-1} \frac{f(y)\, r(x)}{\bar{F}(z)}.$$

The marginal pdf of the nth lower record value can be derived by using the same procedure as that of the nth upper record value. Let $H(u) = -lnF(u), 0 < F(u) < 1$ and $h(u) = -dH(u)/du$, then

$$P(X_{L(n)} \leq x) = \int_{-\infty}^{x} \frac{\{H(u)\}^{n-1}}{(n-1)!} dF(u)$$

and the corresponding pdf $f_{(n)}$ can be written as

$$f_{(n)}(x) = \frac{(H(x))^{n-1}}{(n-1)!} f(x).$$

The joint pdf of $X_{L(1)}, X_{L(2)}, \ldots, X_{L(m)}$ can be written as

$$f_{(1),(2),\ldots,(m)}(x_1, x_2, \ldots, x_m) = h(x_1)h(x_2)\ldots h(x_{m-1})f(x_m), \quad -\infty < x_m < \cdots < x_1 < \infty.$$

The joint pdf of $X_{L(r)}$ and $X_{L(s)}$ is

$$f_{(r),(s)}(x,y) = \frac{(H(x))^{r-1}}{(r-1)!} \frac{[H(y) - H(x)]^{s-r-1}}{(s-r-1)!} h(x) f(y) \tag{2.1}$$

for $s > r$ and $-\infty < y < x < \infty$. Using the transformations $U = H(y)$ and $W = H(x)/H(y)$ in (2.1), it can be shown easily that W is distributed as $B(r, s-r)$. Proceeding as in the case of upper record values, we can obtain the conditional pdfs of the lower record values.

Example 1. Let us consider the exponential distribution with pdf $f(x) = e^{-x}$ and cdf $F(x) = 1 - e^{-x}$, $0 \leq x < \infty$. Then $R(x) = x$ and

$$f_n(x) = \frac{x^{n-1}}{\Gamma(n)} e^{-x}, \qquad x \geq 0$$

The joint pdf of $X_{U(m)}$ and $X_{U(n)}$, $n > m$ is

$$f_{m,n}(x,y) = \frac{x^{m-1}}{\Gamma(m)\Gamma(n-m)} (y-x)^{n-m-1} e^{-y}, \qquad 0 \leq x < y < \infty.$$

The conditional pdf of $X_{U(n)}|X_{U(m)}=x)$ is

$$f(y|X_{U(m)}=x)=\frac{(y-x)^{n-m-1}}{\Gamma(n-m)}e^{-(y-x)}, \qquad 0\leq x<y<\infty.$$

Thus the conditional distribution of $X_{U(n)}$- $X_{U(m)}$ given $X_{U(m)}$ is the same as the unconditional distribution of $X_{U(n-m)}$ for n > m.

Example 2. Suppose that the random variable X has the Gumbel distribution with pdf $f(x)=\exp\{-x\}\exp\{-e^{-x}\}$, $-\infty<x<\infty$. Let $F_{(n)}$ and $f_{(n)}$ be the cdf and pdf of $X_{L(n)}$. It is easy to see that

$$F_{(n)}(x)=\int_{-\infty}^{x}\frac{e^{-nu}}{\Gamma(n)}e^{-e^{-u}}du \quad \text{and} \quad f_{(n)}(x)=\frac{e^{-nx}}{\Gamma(n)}e^{-e^{-x}}, \qquad -\infty<x<\infty.$$

Let $f_{(m,n)}(x,y)$ be the joint pdf of $X_{L(m)}$ and $X_{L(n)}$, m< n. Using (2.1), we get for the Gumbel distribution

$$f_{(m,n)}(x,y)=\frac{(e^{-y}-e^{-x})^{n-m-1}}{\Gamma(n-m)}\frac{e^{-mx}}{\Gamma(m)}e^{-y}e^{-e^{-y}}, \qquad -\infty<y<x<\infty.$$

Thus the conditional pdf $f_{(n|m)}(y|x)$ of $X_{L(n)}|\ X_{L(m)}$ = x is given by

$$f_{(n|m)}(y|x)=\frac{(e^{-y}-e^{-x})^{n-m-1}}{\Gamma(n-m)}e^{-y}e^{-(e^{-y}-e^{-x})}, \qquad -\infty<y<x<\infty.$$

3 Distribution of Inter-record Times

Let $\Delta_r=U(r+1)-U(r)$ and $\Delta_{(r)}=L(r+1)-L(r)$, $r=1,2,\ldots$. We will call Δ_r and $\Delta_{(r)}$ as the upper and lower inter record times respectively. Since $U(1)=1=L(1)$, we have $U(r+1)=1+\Delta_1+\Delta_2+\cdots+\Delta_r$. Similarly $L(r+1)=1+\Delta_{(1)}+\cdots+\Delta_{(r)}$.

Lemma 1. *For any* $n\geq 1$

$$P(\Delta_n<\infty)=1=P(\Delta_{(n)}<\infty).$$

Proof.

$$\begin{aligned}
P(\Delta_1=\infty) &= P\left(\bigcap_{i=2}^{\infty}X_i\leq X_1\right)\\
&= \lim_{m\to\infty}P\left(\bigcap_{i=2}^{\infty}X_i\leq X_1\right)\\
&= \lim_{m\to\infty}P(\max\{X_2,X_3,\ldots,X_m\}\leq X_1)\\
&= \lim_{m\to\infty}\frac{1}{m}=0
\end{aligned}$$

It is obvious that $P(\Delta_{(1)}=\infty)=0$. Similarly it can be shown that $P(\Delta_n=\infty)=P(\Delta_{(n)}=\infty)=0$ for $n\geq 2$. □

Theorem 1.

$$P(\Delta_n = k) = \sum_{i=0}^{k-1} \binom{k-1}{i} (-1)^i (2+i)^{-n}.$$

Proof.

$$\begin{aligned} P(\Delta_1 = k|X_1 = x_1) &= P(U(2) = k+1|X_1 = x_1) \\ &= (F(x_1))^{k-1}\bar{F}(x_1) \end{aligned}$$

Thus

$$\begin{aligned} P(\Delta_1 = k) &= \int_{-\infty}^{\infty} (F(x_1))^{k-1}(1-F(x_1))f(x_1)dx_1 \\ &= \frac{1}{k} - \frac{1}{k+1} \\ &= \frac{1}{k(k+1)} \end{aligned}$$

Similarly it can be shown that

$$P(\Delta_n = k|X_{U(n)} = x_n) = (F(x_n))^{k-1}(1 - F(x_n))$$

$$\begin{aligned} P(\Delta_n = k) &= \int_{-\infty}^{\infty} (F(x))^{k-1}(1-F(x)) \frac{(-\ln(1-F(x))^{n-1}}{\Gamma(n)} f(x)\,dx \\ &= \sum_{i=0}^{k-1} (-1)^i \binom{k-1}{i} \frac{1}{(2+i)^n} \end{aligned}$$

$$\begin{aligned} P(\Delta_n > k) &= \int_{-\infty}^{\infty} \sum_{j=k}^{\infty} (F(x))^{j-1}(1-F(x)) \frac{\{-\ln(1-F(x))\}^{n-1}}{\Gamma(n)} f(x)\,dx \\ &= \sum_{i=0}^{k-1} \binom{k-1}{i} (-1)^i \frac{1}{(1+i)^n}. \end{aligned}$$

Proceeding similarly , the following theorem can be proved. □

Theorem 2.

$$P(\Delta_{(n)} = k) = \sum_{i=0}^{k-1} \binom{k-1}{i} (-1)^i (2+i)^{-n}.$$

Therefore, for all n ≥ 1, Δ_n and $\Delta_{(n)}$ are identically distributed.

Theorem 3.

$$P(\Delta_n \geq s) = \frac{1}{s} \sum_{j=1}^{s} P(\Delta_{n-1} \geq j)$$

Proof.

$$
\begin{aligned}
&P(\Delta_n \geq s)\\
&= \frac{1}{\Gamma(n)} \int_{-\infty}^{\infty} (1-F(x))\,\{-\ln(1-F(x))\}^{n-1} \sum_{k=s}^{\infty} (F(x))^{k-1} f(x)\,dx\\
&= \frac{1}{\Gamma(n)} \int_{-\infty}^{\infty} (F(x))^{s-1}\,\{-\ln(1-F(x))\}^{n-1} f(x)\,dx\\
&= \frac{1}{\Gamma(n)} \int_{0}^{1} t^{s-1}\,\{-\ln(1-t)\}^{n-1}\,dt\\
&= \frac{1}{s\,\Gamma(n-1)} \int_{0}^{1} \frac{1-t^s}{1-t}\,\{-\ln(1-t)\}^{n-2}\,dt\\
&= \frac{1}{s}\sum_{j=1}^{s} P(\Delta_{n-1} \geq j) \qquad (n \geq 2).
\end{aligned}
$$

In particular,

$$
P(\Delta_1 > s) = \frac{1}{s+1} \quad \text{and} \quad P(\Delta_2 > s) = \frac{1}{s+1}\sum_{j=1}^{s}(1+j)^{-1}.
$$

Let $\mathrm{P}_g(\Delta_n(\mathrm{s}))$ be the probability generating function of Δ_n, then

$$
P_g(\Delta_1(t)) = \sum_{x=1}^{\infty} P(\Delta_1 = x)t^x = \sum_{x=1}^{\infty} \frac{1}{x(x+1)}t^x = 1 + \frac{1-t}{t}\ln(1-t),
$$

$$
E[t^{\Delta_n}|X_{U(n)} = x_n] = \sum_{k=1}^{\infty} t^k (F(x_n))^{k-1}(1-F(x_n)) = t\frac{1-F(x_n)}{1-tF(x_n)}
$$

and

$$
\begin{aligned}
P_g(\Delta_n(t)) &= \int_{-\infty}^{\infty} t\frac{1-F(x)}{1-tF(x)}\frac{(-\ln(1-F(x))^{n-1}}{\Gamma(n)} f(x)dx\\
&= \int_{0}^{1} t\frac{1-u}{1-tu}\frac{(-\ln(1-u))^{n-1}}{\Gamma(n)}du\\
&= 1 - \sum_{j=0}^{\infty}(-1)^j \frac{t^j}{(1-t)^j}\cdot\frac{1}{(1+j)^n}, \quad 0 < |t| < 1.
\end{aligned}
$$

We have

$$
P(\Delta_n < m) = 1 - \sum_{j=0}^{m-1}\binom{m-1}{j}(-1)^j \frac{1}{(1+j)^n}.
$$

□

Theorem 4. $E(\Delta_n^{\alpha}) = \infty$ *for* $\alpha \geq 1$ *and* $E(\Delta_n^{\alpha}) < \infty$ *for* $0 < \alpha < 1$.

Proof.

$$\begin{aligned}E(\Delta_n) &= \sum_{s=0}^{\infty} P(\Delta_n > s)\\ &= \int_0^{\infty} \frac{y^{n-1}e^{-y}}{\Gamma(n)} \sum_{s=0}^{\infty} (1-e^{-y})^s\, dy\\ &= \int_0^{\infty} \frac{y^{n-1}}{\Gamma(n)}\, dy\\ &= \infty\end{aligned}$$

and thus $E(\Delta_n^{\alpha}) > E(\Delta_n) = \infty$ for $\alpha > 1$. For $0 < \alpha < 1$,

$$\begin{aligned}E(\Delta_n^{\alpha}) &= \frac{1}{\Gamma(n)} \int_{-\infty}^{\infty} \sum_{k=1}^{\infty} k^{\alpha}(F(x))^{k-1}\bar{F}(x)\{-\ln\bar{F}(x)\}^{n-1} f(x)\, dx\\ &= \frac{1}{\Gamma(n)} \int_0^{\infty} e^{-2t}\, t^{n-1} \sum_{k=1}^{\infty} k^{\alpha}\,(1-e^{-t})^{k-1}\, dt.\end{aligned}$$

Now

$$\begin{aligned}\sum_{k=1}^{\infty} k^{\alpha}\,(1-e^{-t})^{k-1} &\leq \int_0^{\infty} (s+1)^{\alpha}\, e^{-s\ln(1-e^{-t})^{-1}} ds\\ &= \int_1^{\infty} u^{\alpha}\, e^{-(u-1)\ln(1-e^{-t})^{-1}}\, du\\ &\leq (1-e^{-t})^{-1} \int_0^{\infty} u^{\alpha}\, e^{-u\ln(1-e^{-t})^{-1}}\, du\\ &= \left(1-e^{-t}\right)^{-1} \frac{\Gamma(1+a)}{\{-1n(1-e^{-t})\}^{a+1}}\\ &\leq \left(1-e^{-t}\right)^{-1}\Gamma(1+a)\, e^{(1+a)t}.\end{aligned}$$

Thus

$$E(\Delta_n^{\alpha}) \leq \frac{\Gamma(1+\alpha)}{\Gamma(n)} \int_0^{\infty} (1-e^{-t})^{-1} e^{-(1-\alpha)t} t^{n-1}\, dt.$$

Since $x \leq x(1-e^{-x})^{-1} \leq x+1$ we have

$$\begin{aligned}E(\Delta_n^{\alpha}) &\leq \frac{\Gamma(1+\alpha)}{\Gamma(n)} \int_0^{\infty} e^{-(1-\alpha)x}\,(x+1)\,x^{n-2}\, dx\\ &= \frac{\Gamma(1+\alpha)}{(1-\alpha)^n}\left(1+\frac{1-\alpha}{n}\right).\end{aligned}$$

The probability of the event $(\Delta_1 = k_1 \cap \Delta_2 = k_2)$ is

$$\begin{aligned}&P\left(\bigcup_{i=1}^{2} \Delta_i = k_i\right)\\ &= \iint_{-\infty<x_0<x_1<\infty} (F(x_0))^{k_1-1}(F(x_1))^{k_2-1}\bar{F}(x_0)\bar{F}(x_1)r(x_0)f(x_1)\end{aligned}$$

$$
\begin{aligned}
&= \int_{-\infty}^{\infty}(F(x_0))^{k_1-1}\bar{F}(x_0)r(x_0)\left[\frac{1}{k_2}(1-(F(x_0))^{k_2})-\frac{1}{k_2+1}(1-(F(x_0))^{k_2+1})\right]dx_0 \\
&= \frac{1}{k_1k_2}-\frac{1}{k_2(k_1+k_2)}-\frac{1}{(k_2+1)k_1}+\frac{1}{k_2+1}\frac{1}{k_1+k_2+1} \\
&= \frac{1}{k_1(k_1+k_2)(k_1+k_2+1)}.
\end{aligned}
$$

Using the probability of inter-record times, we can easily obtain the probabilities of various events related to the record times. For example,

$$
\begin{aligned}
P(U(2)=n_1) &= P(\Delta_1=n_1-1)=\frac{1}{(n_1-1)n_1}; \\
P(U(3)=n_2,U(2)=n_1) &= P(\Delta_2=n_2-n_1,\Delta_1=n_1-1)=\frac{1}{(n_1-1)(n_2-1)n_2}; \\
P(U(3)=n_2) &= \frac{1}{(n_2-1)n_2}\sum_{j=2}^{n_2-1}\frac{1}{j-1}; \\
P(U(3)=n_2|U(2)=n_1) &= \frac{n_1}{(n_2-1)n_2}.
\end{aligned}
$$

In general

$$
P(U(m+1)=n_m,\ U(2)=n_1,\ldots,U(m)=n_{m-1})=\frac{1}{n_1-1}\cdots\frac{1}{n_{m-1}-1}\frac{1}{n_m}
$$

and

$$
\begin{aligned}
&P(U(m+1)=n_m|U(2)=n_1,\ldots,U(m)=n_{m-1}) \\
&= \frac{n_{m-1}}{n_m(n_m-1)} \\
&= P(U(m+1)=n_m|U(m)=n_{m-1}).
\end{aligned}
$$

Thus, the sequence $\{U(n+1),n\geq 0\}$ forms a homogeneous Markov chain with transition probability $p_{ij}=i/[j(j-1)]$ for $j>i$ and $p_{ij}=0$ for $j\leq i$. Now

$$
E(U(2))=\sum_{n=2}^{\infty}\frac{n}{n(n-1)}=\infty.
$$

Since $U(2)<U(3)<\cdots$, we have $E(U(m+1))=\infty$, for $m\geq 1$. This result can also be established by noting that U(m+1) = $1+\sum_{i=1}^{m}\Delta_i$ and $1+\sum_{i=1}^{m}\Delta_i$ and $E(\Delta_1)=\infty$.

We can write $P(U(2)=m)=|S(1,m-2)|/m!$, where $S(r,m)=$ coefficient of x^r in the expression for $P(x,m)$, where $P(x,m)=x(x-1)(x-2)\ldots(x-m+1)$. $S(r,m)$ is known as Stirling's numbers of first kind. Since

$$
P(U(2)=k,U(3)=m)=\frac{1}{(k-1)m(m-1)},\quad k=2,\ldots,m-1,\ldots
$$

we obtain

$$
P(U(3)=m)=(1+\cdots+\frac{1}{m-2})\frac{1}{m(m-1)},\quad m=3,\ldots
$$

Using the Stirling's number we have $P(U(3)=m)=|S(2,m-1)|/m!$ and in the general case

$$P(U(n)=m)=|S(n-1,m-1)|/m!, \quad m\geq n.$$

Let $P_{gU(n)}(t)$ be the probability generating function of $U(n)$. We have

$$P_{gU(n)}(t)=E(t^{U(n)})=\sum_{m=k}^{\infty} t^m|S(n-1,m-1)|/m!.$$

Using the relation

$$\sum_{m=k}^{\infty} t^m|S(k,m)|m! = (-\ln(1-t))^k/k!,$$

we get

$$\begin{aligned} P_{gU(n)}(t) &= \int_0^t \frac{(-\ln(1-u))^{n-1}}{(n-1)!}du \\ &= \int_0^{-\ln(1-t)} \frac{v^{n-1}}{(n-1)!}e^{-v}dv \\ &= 1-(1-t)\sum_{j=0}^{n-1}\frac{(-\ln(1-t))^j}{j!}. \end{aligned}$$

Williams (1973) showed the following representation for $U(r)$, $r=1,2,\ldots$, where $U(1)=1, U(n+1)|U(n)\exp(W_n)|+1$, and $W_1,W_2,\ldots$ are i.i.d. random variables from exponential distribution with unit mean and $|x|$ denotes the largest integer contained in x. We have

$$P(\exp(-W_n)<x)=P(W_n>-\ln x)=x \quad (0<x<1),$$

$$P(U(n+1)=j|U(n)=m)=P\Big(\frac{L(n)}{V_n}+1=j|L(n)=m\Big),$$

where the random variable V_n is uniformly distributed on $[0,1]$ and independent from $L(n)$. Thus

$$\begin{aligned} P(U(n+1)=j|U(n)=m) &= P\left(\frac{m}{V_n}+1=j|U(n)=m\right) \\ &= P\left(\frac{m}{V_n}+1=j\right) \\ &= P\left(\frac{i}{m}<V_n<\frac{i}{m-1}\right)=\frac{i}{m(m-1)} \end{aligned}$$

Galambos and Seneta (1975) proved that if

$$T(n)-1<\frac{U(n)}{U(n-1)}\leq T(n), \qquad n=2,3,\ldots,$$

then $T(r)$, $r=2,3,\ldots$ are independent and identically distributed with

$$P(T(r)=j)=P(U(2)=j)=\frac{1}{j(j-1)}, \qquad j=2,3,\ldots$$

□

4 Representation of Records

Let $Y_1, Y_2, \ldots, Y_n, \ldots$ be a sequence of independent and identically distributed random variable with cdf $F_0(x) = 1 - \exp(-x),\ x > 0$. Further suppose that $X_1, X_2, \ldots$ be a sequence i.i.d. r.v.'s with continuous cdf F. Then one has

$$\ln(1 - F(x)) \stackrel{d}{=} Y_1.$$

The following theorem (Bairamov and Ahsanullah (2000)) gives the representation of the nth record as sum of n independent random variables.

Theorem 5.

$$X_{U(n)} \stackrel{d}{=} g_F^{-1}(g_F(X_1) + g_F(X_2) + \cdots + g_F(X_n)),$$

where $g_F(x) = -\ln(1 - F(x)$ *and* $g_F^{-1}(x) = F^{-1}(1 - e^{-x})$.

Using the result of Example 1, we obtain

$$\begin{aligned} & P(X_{U(n)} \leq x) \\ &= P(Y_1 + Y_2 + \ldots + Y_n \\ &= P(-\ln(1 - F(x_1) + (-\ln(1 - F(X_2)) + \ldots + (-\ln(1 - F(x_n)) \\ &= P(g_F(X_1) + g_F(X_2) + \ldots + g_F(X_n)) \\ &= P(g_F^{-1}\{g_F(X_1) + g_F(X_2) + \ldots + g_F(X_n)\}. \end{aligned}$$

Therefore

$$X_{U(n)} \stackrel{d}{=} g_F^{-1}(g_F(X_1) + g_F(X_2) + \cdots + g_F(X_n)). \tag{4.1}$$

Using (4.1) we will give here the representation of the nth record value of several well known distribution (for more details see Ahsanullah (2004)).

Example 3. For the two parameter exponential distribution, $E(\mu, \sigma)$, with cdf $F(x) = 1 - e^{-(x-\mu)/\sigma}$, $x \geq 0$, we obtain

$$X_{U(n)} \stackrel{d}{=} X_1 + X_2 + \cdots + X_n - (n-1)\mu$$

Example 4. Consider the Weibull distribution, $W(\alpha, \beta)$ with $F(x) =$ 1-$e^{-\beta x^\alpha}$ $(x \geq 0, \alpha > 0, \beta > 0)$ Using (4.1), we obtain

$$X_{U(n)} \stackrel{d}{=} (X_1^\alpha + X_2^\alpha + \ldots + X_2^\alpha)^{\frac{1}{\alpha}} \beta^{\frac{1}{2}}.$$

Example 5. Consider the Power Function distribution, POW (δ, α, β) with cdf

$$F(x) = \begin{cases} 0, & x < 0 \\ 1 - \left(\frac{\beta - x}{\beta - \alpha}\right)^\theta, & \alpha \leq x \leq \beta,\ \theta > 0,\ -\infty < \alpha < \beta < \infty, \\ 1, & x \geq \beta. \end{cases}$$

For this distribution, the following representation is true

$$X_{U(n)} \stackrel{d}{=} \beta - (\beta - \alpha)^{\frac{1}{\theta}} (\beta - X_1)(\beta - X_2) \ldots (\beta - X_n).$$

The following theorem can be proved using the same procedure as in the proof of Theorem 5. For details, see Ahsanullah and Nevzorov (2001, p.242).

Theorem 6. *Suppose that* $\{X_1, X_2, \ldots\}$ *and* $\{Y_1, Y_2, \ldots\}$ *are two sequences of i.i.d. r.v.'s with continuous distribution functions* F *and* H*, respectively. Then*

$$(X_{U(1)}, \ldots, X_{U(n)}) \stackrel{d}{=} (G(H(Y_{U(1)})), \ldots, G(H(Y_{U(n)})),$$

where G *is the inverse function of* F.

5 Discrete Distributions

Let $X_1, X_2, \ldots, X_n, \ldots$ be a sequence independent and identically distributed random variables taking values on $0, 1, 2, \ldots$ such that $F(n) < 1$ for all $n = 0, 1, 2, \ldots$ Let $p_k = P(X_1 = k)$, $P(k) = \sum_{j=0}^{k} p(j)$, $k \geq 0$ and $\bar{P}(k) = 1 - P(k)$ with $P(\infty) = 1$. The joint probability mass function (pmf) of $X_{U(1)}, \ldots X_{U(n)}$ is defined for $0 \leq x_1 < \ldots < x_n < \infty$ as

$$\begin{aligned} p_{1,2,\ldots,n}(x_1, x_2, \ldots, x_n) &= P(X_{U(1)} = x_1, X_{U(2)} = x_2, \ldots, X_{U(n)} = x_n) \\ &= \frac{p(x_1)}{\bar{P}(x_1)} \frac{p(x_2)}{\bar{P}(x_2)} \cdots \frac{p(x_{n-1})}{\bar{P}(x_{n-1})} p(x_n). \end{aligned}$$

The marginal pmf's of the upper record values are given as

$$p_1(x_1) = P(X_{U(1)} = x_1) = p(x_1), \quad x_1 = 0, 1, 2, \ldots,$$

$$p_2(x_2) = P(X_{U(2)} = x_2) = R_1(x_2) p(x_2),$$

where

$$R_1(k) = \sum_{0 \leq x_1 < x_2} B(x_1) \quad \text{and} \quad B(x) = \frac{p(x)}{\bar{P}(x)}, \quad x_2 = 1, 2, \ldots$$

Finally,

$$p_n(x_n) = P(X_{U(n)} = x_n) = R_{n-1}(x_n) p(x_n),$$

where

$$R_{n-1}(x_n) = \sum_{0 \leq x_1 < x_2 < \cdots < x_{n-1} < x_n} B(x_1) B(x_2) \ldots B(x_{n-1}), \qquad x_n = n-1, n, \ldots$$

The joint pmf of $X_{U(m)}$ and $X_{U(n)}$, $m < n$ is given by

$$\begin{aligned} p_{m,n}(x_m, x_n) &= P(X_{U(m)} = x_m, X_{U(n)} = x_m) \\ &= R_{m-1}(x_m) A(x_m) R_{m,+1,n}(x_m, x_n) p(x_n), \quad m \leq x_m < x_n - n + m < \infty, \end{aligned}$$

where

$$R_{m+1,n}(x, y) = \sum_{x_m < x_{m+1} < \cdots < x_n} B(x_{m+1}) \ldots B(x_{n-1}), \quad m < n-1 \quad \text{and} \quad R_{n,n}(x, y) = 1.$$

The conditional pmf of $X_{U(n)}$ given $X_{U(m)} = x_m$ is

$$p_{n|m}(X_{U(n)} = x_n | X_{U(m)} = x_m) = R_{m,n}(x_m, x_n) \frac{p(x_n)}{\vec{P}(x_m)}, \quad x_m \leq x_n - n + m < \infty$$

and

$$P_{n|n-1}(X_{U(n)} = x_n | X_{U(n-1)} = x_{n-1}) = \frac{p(x_n)}{\vec{P}(x_{n-1})}, \quad x_{n-1} < x_n. \tag{5.1}$$

It follows that the sequence of upper record values $X_{U(1)}, X_{U(2)} \ldots$ forms a Markov chain. Let $I_n = n$ if n is a record value i.e. $X_{U(m)} = n$, for $m = 1, 2, \ldots$ and $I_n = 0$ if $X_{U(1)}, X_{U(2)}, \ldots$ does not contain the value n.

The following theorem is due to Aliev and Ahsanullah (2002).

Theorem 7. *Let X_1, $X_2, \ldots$ be a sequence of independent and identically distributed random variables taking values on $0, 1, 2, \ldots$ with common distribution function F such that $F(n) < 1$ for all n and $E(X_i^2) < \infty$. Suppose that $\{B_k,\ k = 0, 1, \ldots|$ be a sequence of numbers such that $2 + 2B_{n+1} - B_s - B_{s+2} \geq 0\}$. If there exits $F(x)$ such that $E\{(X_{U(2)} - X_{U(1)})^2 | X_{U(1)}\} = B_s$, $s = 0, 1, 2 \ldots$. Then $F(x)$ is unique.*

Proof. From (5.1), we obtain

$$E[(X_{U(2)} - X_{U(1)})^2 | X_{U(1)} = s] = \frac{\sum_{j=1}^{\infty} j^2 p_{s+j}}{\sum_{j=1}^{\infty} p_{s+j}},$$

where $p_j = P(X = j)$, $j = 0, 1, \ldots$ Thus the condition $E[(X_{U(2)} - X_{U(1)})^2 | X_{U(1)}] = B_s$ implies

$$\sum_{j=1}^{\infty} j^2 p_{s+j} = B_s \sum_{j=1}^{\infty} p_{s+j} \tag{5.2}$$

for $s \geq 0$. Writing $s = s + 1$, we get from (5.2)

$$\sum_{j=1}^{\infty} (j-1)^2 p_{s+j} = B_{s+1} \sum_{j=1}^{\infty} p_{s+j+1} \tag{5.3}$$

Subtracting (5.3) from (5.2) we obtain

$$\sum_{j=1}^{\infty} (2j-1) p_{s+j} = B_s \sum_{j=1}^{\infty} p_{s+j} - B_{s+1} \sum_{j=1}^{\infty} p_{s+1+j} \tag{5.4}$$

Now substituting $s = s + 1$ in (5.4), we will have

$$\sum_{j=1}^{\infty} (2j-1) p_{s+1+j} = B_{s+1} \sum_{j=1}^{\infty} p_{s+1+j} - B_{s+2} \sum_{j=1}^{\infty} p_{s+2+j} \tag{5.5}$$

Subtracting (5.5) from (5.4) and on simplification , we obtain

$$p_{s+2} = \frac{1 + 2B_s + 1 - B_{s+2}}{B_{s+2}} p_{s+1} - \frac{2 + 2B_{s+1} - B_s - B_{s+2}}{B_{s+2}} (1 - p_0 - p_1 - \ldots - p_s) \tag{5.6}$$

Since the coefficients $-(2 + 2B_{s+1} - B_s - B_{s+2})/B_{s+2}$ and $(1 + 2B_{s+1} - B_{s+2})/B_{s+2}$ are positive, it means that p_{s+2} is increasing (decreasing) if p_{s+1} increases(decreases) for all ≥ 0. It means that for any given p_0 all probabilities $p_2, p_3, \ldots$ increase when p_1 increases. Together with the condition $\sum_{i=0}^{\infty} p_i = 1$ we conclude that for any given p_0, we have only one F(x) which satisfies (5.6). □

6 Geometric distribution

A discrete random variable X is said to have geometric distribution if its probability mass function (pmf) is of the following form:

$$p(k) = P(X = k) = pq^{k-1}, \qquad 0 < p < 1, \quad q = 1 - p, \quad k \in A_0, \tag{6.1}$$

where A_n is the set of integers $n+1, n+2, \ldots$ $(n \geq 0)$. We say $X \in$ GE(p), if the pmf of X is as given in (6.1). For k > 0, we define $r(k) = P(X = k|X \geq k)$. We choose to distinguish between GE(p) and the larger class of distributions having geometric tail (GET). We write $X \in$ GET(s, p) if the pmf of X is as follows:

$$p(k) = P(X = k) = cq^{k-1}, \qquad q = 1 - p, \quad k \in A_s,$$

where c is such that $\sum_{k=s+1}^{\infty} p(k) = 1$. If $s = 0$, then GET(s, p) = GE(p) with $c = p$.

The geometric distribution like the exponential distribution possesses the memory less property, i.e.

$$\bar{P}(r+s) = \bar{P}(r)\bar{P}(s),$$

where r and s are positive integers and $\bar{P}(j) = \sum_{k=j+1}^{\infty} p(k)$. Geometric distribution is said to be a discrete analogue of the exponential distribution.

If $X \in$ GE(p), then $\bar{P}(x) = q^x$ and $p(x) = pq^{x-1}$, for $x \in A_0$. Substituting the values of $\bar{P}(x_i)$ and $p(x_i)$ in (6.1), we get

$$p(x_1, x_2, \ldots, x_m) = p^n q^{x_m - m}, \quad 1 \leq x_1 < x_2 < \cdots < x_m < \infty.$$

The conditional pmf of $X_{U(n)}|X_{U(n-1)} = x_{n-1}$ is

$$P(X_{U(n)} = x_n|X_{U(n-1)} = x_{n-1}) = pq^{x_n - x_{n-1} - 1}, \quad n - 1 \leq x_{n-1} < x_n < \infty.$$

Thus $X_{U(n)} - X_{U(n-1)}$ is independent of $X_{U(n-1)}$ and $X_{U(n)} - X_{U(n-1)} \in GE(p)$, $n = 2, 3, \ldots$. Let $V_1 = X_{U(1)}$ $V_2 = X_{U(2)} - X_{U(1)}$ $V_n = X_{U(n)} - X_{U(n-1)}$. Then V_i's are independent and $V_i \in GE(p)$. We have

$$X_{U(n)} = V_1 + V_2 + \cdots + V_n. \tag{6.2}$$

It is known that if $X \in GE(p)$, then

$$E(s)^X = \sum_{x=1}^{\infty} s^x pq^{x-1} = \frac{ps}{1-qs}.$$

Using (6.2), we obtain

$$E(s^{X_{U(n)}}) = E(s^{V_1+V_2+\cdots+V_n}) = \left(\frac{ps}{1-qs}\right)^n.$$

The coefficient of s^x in $(ps/(1-qs))^n$ is $\binom{x-1}{n-1} p^n q^{x-n}$ $(x \geq n)$. Thus the marginal pmf of $X_{U(m)}$ can be written as

$$p_m(x) = P(X_{U(m)} = x) = \binom{x-1}{m-1} p^m q^{x-m}, \quad x \in A_{m-1}, \quad m \geq 1.$$

We see that $X_{U(m)}$has a negative binomial distribution with parameters m and p. We can write for $n > m$

$$X_{U(n)}|X_{U(m)} = x_m \stackrel{d}{=} U_{m+1} + \cdots + U_n + x_m$$

and

$$E\left(s^{X_{U(n)}}|X_{U_{(m)}} = x_m\right) = s^{x_m}\left(\frac{ps}{1-qs}\right)^{n-m}.$$

Considering the coefficient of s^y in $s^{x_n}\left(\frac{ps}{1-qs}\right)^{n-m}$

$$p_{n|m}(y|x) = P(X_{U(n)} = y|X_{U(m)} = x) = \binom{y-x-1}{n-m-1} p^{n-m} q^{y-x-n+m}$$

we obtain the conditional pmf of $X_{U(n)}$given $X_{U(m)}$ as

$$P(X_{U(n)} = y|X_{U(m)} = x) = \binom{y-x-1}{n-m-1} p^{n-m} q^{y-x-n+m}, \quad 0 < m \leq x \leq y-n+m < \infty.$$

But we know that the marginal pmf of $X_{U(m)}$ is

$$P(X_{U(m)} = x) = \binom{x-1}{m-1} p^m q^{x-m}, \quad x \in A_{m-1}, \quad m \geq 1.$$

Thus, the joint pmf of $X_{U(m)}$and $X_{U(n)}$ is

$$P(X_{U(m)} = x, X_{U(n)} = y) = \binom{x-1}{m-1}\binom{y-x-1}{n-m-1} p^n q^{y-n}, \quad m \leq x < y-n+m < \infty.$$

Let $Z_{m,n} = X_{U(n)} - X_{U(m)}$, 0 < m < n < ∞, then

$$\begin{aligned} P(Z_{m,n} = z|X_{U(m)} = x) &= P(X_{U(n)} = z+x|X_{U(m)} = x) \\ &= \binom{z-1}{n-m-1} p^{n-m}(1-p)^{z-n+m}, \quad z \in A_{n-m-1}. \end{aligned}$$

Thus $Z_{n,m}$ and $X_{U(m)}$ are independent. Further $Z_{n,m}$ and $X_{U(n-m)}$ are identically distributed.

7 Characterizations

Here we present some characterizations of discrete distributions. There are several characterizations of the geometric distributions based on (i) independence of $X_{U(n)} - X_{U(m)}$ and $X_{U(m)}$ (ii) conditional distribution of $X_{U(n)}|X_{U(m)}$, and (iii) moment properties of some functions of $X_{U(n)}$ for $n > m$.

The following Theorem is proved by Srivastava (1979).

Theorem 8. *Suppose $F(x)$ is the distribution function of the sequence of i.i.d. random variables $\{X_n, n \geq 1\}$ with positive mass function only at $1, 2, \ldots$ Then $P(X_{U(2)} - X_{U(1)} = 1|X_{U(1)} = i) = P(X_{U(2)} - X_{U(1)} = 1)$ for $i = 1, 2, \ldots,$ if and only if X_n has the geometric distribution with pmf given by*

$$p_j = P(X = j) = cp(1-p)^{j-2}, \quad j = 2, 3, \ldots \tag{7.1}$$

and

$$p_1 = 1 - \sum_{j=2}^{\infty} p_j = 1 - c, \quad 0 < p < 1, \quad 0 < c < 1.$$

Proof. We give here the original proof of Srivastava.

$$\begin{aligned} & P(X_{U(2)} - X_{U(1)} = 1 | X_{U(1)} = i) \\ = & \frac{P(X_{U(2)} = i+1, X_{U(1)} = i)}{P(X_{U(1)} = i)} \\ = & \frac{p_{i+1}}{1-(p_1 + \cdots + p_i)}. \end{aligned}$$

Since this conditional probability is independent of i, we must have

$$p = \frac{p_{i+1}}{1-(p_1 + \cdots + p_i)}, \qquad j = 1, 2, \ldots$$

For $j = 1$, we get $p_2 = p(1-p_1) = cp$, say. For $j = 2$, we get $p_3 = p(1-p_1-p_2) = (1-p_1)p(1-p) = cp(1-p)$. Similarly for $j = k$, we have

$$p_k = cp(1-p)^{j-2}, \qquad k \geq 3. \tag{7.2}$$

If p_j is given by (7.1), then

$$P(X_{U(2)} - X_{U(1)} = 1 | X_{U(1)} = i) = P(X_{U(2)} - X_{U(1)} = 1).$$

A generalization of Theorem 8 is the following theorem. □

Theorem 9. *Let $\{X_n,\ n \geq 1\}$ be a sequence of independent and identically distributed discrete random variables with common distribution function F. Suppose X_n is concentrated on the positive integers and $a = \sup\{x | F(x) < 1\} = \infty$. Then $X_n \in GET(n, p)$ for some fixed $n \geq 1$, if and only if $X_{U(n+1)} - X_{U(n)}$ and $X_{U(n)}$ are independent.*

Proof. The 'only if' part follows immediately from equation (7.2), so we need to established the 'if' part. For $x \in A_0$. Let $c(u) = P(X_{U(n+1)} - X_{U(n)} = u | X_{U(n)} = x)$, then

$$\begin{aligned} c(u) &= \frac{p(u+x)}{\bar{P}(x)} \qquad (7.3) \\ &= \frac{\bar{P}(u+x-1) - \bar{P}(u+x)}{\bar{P}(x)}, \qquad u \in A_0, \quad x \in A_n. \end{aligned}$$

Summing both sides of (7.3) with respect to u from 1 to u_0, and writing

$$c_1(u_0) = \sum_{u=1}^{u_0} c(u) \quad \text{and} \quad c_0 = 1 - c_1(u_0), \tag{7.4}$$

on simplification, we get

$$\bar{P}(x + u_0) = c_0(u_0)\bar{P}(x), \quad u \in A_0, \quad x \in A_n. \tag{7.5}$$

The general solution of (7.5) is

$$\bar{P}(x) = cp^x, \quad x \in A_n,$$

where c is independent of p. Using the boundary condition $\bar{P}(\infty) = 0$, we get $\bar{P}(x) = cp^x$, $x \in A_n$, $0 < q < 1, x \in A_n$.

We have already seen that for $u \in A_0$

$$P(X_{U(n+1)} - X_{U(n)} = u | X_{U(n)} = y) = pq^{u-1} = P(X_k = u).$$

Does the above condition characterize the geometric distribution? As an answer to that question we have the following theorem. □

Theorem 10. *Let $\{X_n,\ n \geq 1\}$ be a sequence of independent and identically distributed random variables with common distribution function F. Suppose X_n is concentrated on the positive integers with $a = \sup\{x|F(x) < 1\} = \infty$. Further if $P(X_{U(n+1)} - X_{U(n)} = u|X_{U(n)} = y) = P(X_1 = u)$ for two fixed $y \in A_{n-1}$ and y_1 and y_2 are relatively prime and all $u \in A_0$, then $X_1 \in GET(n,p)$.*

Proof. Suppose that $P(X_{U(n+1)} - X_{U(n)} = u|X_{U(n)} = y) = P(X_1 = u)$, then from (7.4), we have

$$P(X_{U(n+1)} - X_{U(n)} = u|X_{U(n)} = y) = \frac{p(u+y)}{\bar{P}(y)} = p(u) \tag{7.6}$$

for two relatively prime $y_1, y_2 \in A_{n-1}$ and all $u \in A_0$. Summing (7.6) with respect to u from $u_0 + 1$ to ∞, we get

$$\frac{\bar{P}(u_0 + y)}{\bar{P}(y)} = \bar{P}(u_0) \tag{7.7}$$

for two relatively prime $y_1,\ y_2 \in A_{n-1}$ and all $u \in A_0$. The general solution of (7.7) is

$$\bar{P}(x) = cp^x, \qquad x \in A_n,$$

and since $\bar{P}(\infty) = 0$, we must have

$$\bar{P}(x) = cp^x, \qquad 0 < q < 1, \quad x \in A_n.$$

Srivastava (1979) gave a characterization of the geometric distribution using the condition $E(X_{U(2)}|X_{U(1)} = y) = \alpha + y$. Ahsanullah and Holland (1984) proved the following theorem which is a generalization of Srivastava's result. □

Theorem 11. *Let $\{X_n,\ n \geq 1\}$ be a sequence of independent and identically distributed discrete random variables with common distribution function F. Suppose X_n is concentrated on the positive integers with $a = \sup\{x|F(x) < 1\} = \infty$. Further suppose $E(X_{U(n+1)})^2 < \infty$. If $E(X_{U(n+1)}|X_{U(n)} = y) = y + p^{-1}$ for all $y \in A_{n-1}$, then $X_1 \in GET(n,p)$ and $0 < p < 1$.*

Proof. Suppose

$$E(X_{U(n+1)}|X_{U(n)} = y) = p^{-1} + y, \quad \text{for all } y \in A_{n-1}.$$

We have that for $y \in A_{n-1}$

$$P(X_{U(n+1)} = u + y | X_{U(n)} = y) = \frac{p(u+y)}{\bar{P}(y)}$$

and hence

$$\begin{aligned} E(X_{U(n+1)} | X_{U(n)} = y) &= \sum_{u=y}^{\infty} (u+y) \frac{p(u+y)}{\bar{P}(y)} \\ &= p^{-1} + y, \end{aligned}$$

which implies that

$$\sum_{u=y}^{\infty} u \frac{p(u+y)}{\bar{P}(y)} = p^{-1}.$$

Thus, we have

$$\begin{aligned} p^{-1} &= \frac{\sum_{u=1}^{\infty} u[\bar{P}(u+y-1) - \bar{P}(u+y)]}{\bar{P}(y)} \\ &= \sum_{i=0}^{\infty} \frac{\bar{P}(y+i)}{\bar{P}(y)} \\ &= \frac{J(y)}{J(y) - J(y+1)}, \end{aligned}$$

where $J(y) = \sum_{i=0}^{\infty} \bar{P}(y+i)$. Thus, for all $y \in A_{n-1}$

$$J(y+1) = qJ(y), \quad q = 1 - p. \tag{7.8}$$

The general solution of (7.8) with the boundary condition, $J(n) = \theta$, $0 < \theta < 1$, is

$$J(x) = \theta q^{x-n}, \qquad x \in A_{n-1}.$$

Therefore

$$\bar{P}(x) = J(x) - J(x+1) = \theta p q^{x-n}, \qquad x \in A_{n-1}$$

i.e. $X_n \in GET(n, p)$.

We have seen that for $u \in A_0$

$$P(X_{U(n+1)} - X_{U(n)} = u | X_{U(n)} = y) = pq^{u-1} = P(X_k = u).$$

Thus,

$$\begin{aligned} & E((X_{U(n+1)} - X_{U(n)})^2 | X_{U(n)} = y) \\ &= \sum_{u=1}^{\infty} u^2 p q^{u-1} \\ &= \frac{1+q}{p^2}. \end{aligned}$$

Thus, $E((X_{U(n+1)} - X_{U(n)})^2 | X_{U(n)} = y)$ is independent of y. If fact, it can be shown that $E((X_{U(n+1)} - X_{U(n)})^k | X_{U(n)} = y)$, for any positive integer k, is independent of y. □

8 Weak Records

Vervaat (1973) introduced the concept of weak records of discrete distribution. For simplicity we will assume that the random variables X_i, $i = 1,2,\ldots$ are concentrated on the non-negative integers. If in the definition of record times and record values we replace $>$ by $\geq$, then we obtain weak record times and weak record values. The joint pmf of $X_{U_w(1)}, X_{U_w(2)}, \cdots, X_{U_w(n)}$ is given by

$$p_{w,1,2,\ldots,n}(x_1,x_2,\ldots,x_n) = \left(\prod_{i=1}^{n-1} \frac{p(x_i)}{\bar{P}(x_i-1)}\right) p(x_n), \quad 0 \leq x_1 \leq \ldots \leq x_n < \infty. \tag{8.1}$$

For any $m > 1$ and $n > m$, we can write

$$\begin{aligned} P\left(X_{U_w(n)} = x_n, \ldots, X_{U_w(m+1)} = x_{m+1} | X_{U_w(m)} = x_m, \ldots, X_{U_w(1)} = x_1\right) \\ = \left(\prod_{i=m+1}^{n-1} \frac{p(x_i)}{\bar{P}(x_i-1)}\right) \frac{p(x_n)}{\bar{P}(x_m-1)}. \end{aligned} \tag{8.2}$$

It follows easily from (8.1)and (8.2) that the weak records, $X_{U_w(1)}, X_{U_w(2)}, \ldots$ form a Markov chain.

The marginal pmf's of the upper weak records are given by

$$P(X_{U_w(1)} = x_1) = p_{w,1}(x_1) = p(x_1), \quad x_1 = 0,1,\ldots,$$

$$P(X_{U_w(2)} = x_2) = p_{w,2}(x_2) = R_{w,1}(x_2)p(x_2), \quad x_2 = 0,1,\ldots$$

where

$$R_{w,1}(x_2) = \sum_{0 \leq x_1 \leq x_2} \frac{p(x_1)}{\bar{P}(x_1-1)}.$$

Finally,

$$P(X_{U_w(n)} = x_n) = p_{w,n}(x_n) = R_{w,n-1}(x_n)p(x_n),$$

where

$$R_{w,n-1}(x_n) = \sum_{0 \leq x_1 \leq x_2 \leq \cdots x_{n-1}} \prod_{i=1}^{n-1} \frac{p(x_i)}{\bar{P}(x_i-1)}.$$

The joint pmf of $X_{U_w(m)}$ and $X_{U_w(n)}$ for $m < n$ is given by

$$P_{w,m,n}(x_m,x_n) = R_{w,m}(x_m)A_w(x_m)R_{w,m+1,n}(x_m,x_n)p(x_n), \quad m \leq x_m < x_n - n + m < \infty,$$

where

$$R_{w,m,n}(x,y) = \sum_{x_m \leq x_{m+1} \leq \cdots \leq n} A_w(x_{m+1})\ldots A_w(x_{n-1}), \quad m < n-1,$$

$R_{w,n-1,n}(x,y) = 1$ and $A_w(x) = p(x)/\bar{P}(x-1)$. The conditional pmf of $X_{U_w(m)}$ given $X_{U_w(n)}$ is

$$P_{w,n|m}(X_{U_w(n)} = x_n | X_{U_w(m)} = x_m) = R_{w,m+1,n}(x_m,x_n)\frac{p(x_n)}{\bar{P}(x_m-1)},$$

for $m \le x_m \le x_n < \infty$. Thus, the pmf of $X_{U_w(n)}$ given $X_{U_w(n-1)}$ is

$$P_{w,n|n-1}(X_{U_w(n)} = x_n | X_{U_w(n-1)} = x_{n-1} = \frac{p(x_n)}{\bar{P}(x_{n-1}-1)} \tag{8.3}$$

and

$$P_{w,n|n-2}(X_{U_w(n)} = x_n | X_{U_w(n-1)} = x_{n-2}) = R_{w,n-2,n}(x_{n-2}, x_n) \frac{p(x_n)}{\bar{P}(x_{n-2}-1)}.$$

Theorem 12. *Let $\{X_i,\ i = 1,2,\ldots\}$ be sequence of independent and identically distributed random variables taking values on $0,1,\ldots,n \le \infty$ with distribution F such that $F(n) < \infty$ for $n < \infty$ and $E(X_1 \ln(1+X_1) < \infty$. For some continuous function ψ, the condition $E(\psi(X_{U_w(n)}|X_{U_w(n-1)} = j) = g(j)$ determines the distribution.*

Proof. We have from (8.3)

$$P(\psi(X_{U_w(n)} = y)|X_{U_w(n-1)} = x) = \frac{p_y}{q_x},$$

where $p_y = P(X = y)$ and $q_x = P(X \ge x)$. Now

$$E(\psi(X_{U_w(n)}|X_{U_w(n-1)} = j) = \frac{1}{q_j}\sum_{k=j}^{N} \psi(k) p_k.$$

Using the condition as given in the theorem, we can write the above expression as

$$g(j)q_j = \sum_{k=j}^{n} \psi(k) p_k \tag{8.4}$$

Taking first order difference, we obtain from (8.4)

$$g(j)q_j - g(j+1)q_{j+1} = \psi(j) p_j.$$

Thus,

$$g(j)q_j - g(j+1)(q_j - p_j) = \psi(j) p_j \tag{8.5}$$

i.e.

$$p_j = \frac{g(j+1) - g(j)}{g(j+1) - \psi(j)} q_j.$$

Since $q_0 = 1$ and

$$q_j = \frac{q_j}{q_{j-1}} \frac{q_{j-1}}{q_{j-2}} \cdots \frac{q_1}{q_0},$$

we have

$$p_j = \frac{g(j+1) - g(j)}{g(j+1) - \psi(j)} \prod_{k=0}^{j-1} \left(\frac{q_{k+1}}{q_k} \right). \tag{8.6}$$

From (8.5)

$$g(j)q_j - g(j+1)q_{j+1} = \psi(j)(q_j - q_{j+1})$$

i.e.

$$\frac{q_{j+1}}{q_j} = \frac{g(j) - \psi(j)}{g(j+1) - \psi(j)}$$

From (8.6) we can write

$$p_j = \frac{g(j+1)-g(j)}{g(j+1)-\psi(j)} \prod_{k=0}^{j-1} \left(\frac{g(k)-\psi(k)}{g(k+1)-\psi(k)} \right). \qquad \square$$

Example 6. (Geometric distribution) Let $X_1, X_2, \ldots$ be a sequence of independent and identically random variables with $p(k) = pq^k$ and $\bar{P}(k-1) = q^k$, $k = 0,1,2,\ldots$. Here $R_{w,1}(x_2) = \sum_{0 \le x_1 \le x_2} p(x_1)/\bar{P}(x_1-1) = x_{2+1}p$. Thus, $p_{w,2}(k) = R_{w,1}(k)p(k) = (k+1)p^2q^k$, $k = 0,1,\ldots$ Since

$$P_{w,n|n-1}(X_{U_w(n)} = x_n | X_{U_w(n-1)} = x_{n-1}) = \frac{p(x_n)}{\bar{P}(x_{n-1}-1)} = pq^{x_n - x_{n-1}}, \qquad x_n \ge x_{n-1},$$

we obtain for $x_3 = 0,1,\ldots$

$$\begin{aligned} p_{w,3}(x_3) &= \sum_{x_2=0}^{x_3} (x_2+1)p^2q^{x_2}pq^{x_3-x_2} \\ &= \sum_{x_2=0}^{x_3} (x_2+1)p^3q^{x_3} \\ &= \frac{(x_3+2)(x_3+1)}{2}p^3q^{x_3}. \end{aligned}$$

By induction it can be proved that when $n \ge 2$ and $x_n = 0,1,2,\ldots$

$$\begin{aligned} p_{w,n}(x_n) &= \frac{(x_n+1)(x_n+2)\ldots(x_n+n-1)}{(n-1)!}p^nq^{x_n} \\ &= \binom{x_n+n-1}{n-1}p^nq^{x_n} \end{aligned}$$

and

$$\begin{aligned} E(X_{U_w(n)} | X_{U_w(n-1)} = x_{n-1}) &= \sum_{x_n = x_{n-1}}^{\infty} x_n pq^{x_n - x_{n-1}} \\ &= x_{n-1} + \frac{q}{p}. \end{aligned}$$

The conditional pmf of $X_{U_w(n)}$ given $X_{U_w(n-2)}$ is for $0 \le x_{n-2} \le x_n < \infty$

$$P_{w,n|n-2}(X_{U_w(n)} = x_n | X_{U(n-2)} = x_{n-2}) = R_{w,n-2,n}(x_n, x_{n-2}) \frac{p(x_n)}{\bar{P}(x_{n-2}-1)},$$

where

$$R_{w,n-2,n}(x,y) = \sum_{x_{n-2} \le x_{n-1} \le x} A_w(x_{n-1}) = x - x_{n-2} + 1.$$

The conditional expectation of $X_{U_w(n)}|X_{U_w(n-2)} = x$ is

$$\begin{aligned} & E(X_{U_w(n)}|X_{U_w(n-2)} = x_{n-2}) \\ = & \sum_{x=x_{n-2}}^{\infty} x(x-x_{n-2}+1)p^2q^{x-x_{n-2}} \\ = & x_{n-2} + \sum_{x=x_{n-2}}^{\infty} (x-x_{n-2})(x-x_{n-2}+1)p^2q^{x-x_{n-2}} \\ = & x_{n-2} + \frac{2q}{p}. \end{aligned}$$

Stepanov (1994) and Aliev (1998) proved that if $E(X_{U_w(n)}|X_{U_w(n-1)} = x_{n-1}) = x_{n-1} + b$ then X_i's $\in GE(1/(1+b))$. This result follows from equation (8.4) if we take $\psi(x) = x$ and $g(x) = x+b$. A generalization is the following theorem due to Wesolowski and Ahsanullah (2001).

Theorem 13. *Let $X_1, X_2, \ldots$ be a sequence of independent and identically distributed random variables taking values on $0,1,\ldots,N$ with distribution function F such that $F(n) < 1$ for any $n < N$ and N need not be finite. If $E(X_{U_w(n)}|X_{U_w(n-2)} = x) = x+b$, $n > 2$, then X_i's $\in GE(b/(2+b))$.*

Proof. Let $p_k = P(X = k)$ and $q_k = p(X \geq k)$. Then by assumption

$$E(X_{U_w(n)}|X_{U_w(n-2)} = m) = \frac{1}{q_m}\sum_{r=m}^{N} \frac{p_r}{q_r}\sum_{k=r}^{N} kp_k = m+b.$$

We can write

$$\sum_{r=m}^{N} \frac{p_r}{q_r}\sum_{k=r}^{N} kp_k = (m+b)q_m. \tag{8.7}$$

On taking first order differences, we obtain from (8.7) for $m = 0,1,\ldots,N$

$$(m+b)\frac{q_m^2}{p_m} - (m+1+b)\frac{q_m q_{m+1}}{p_m} = \sum_{k=m}^{N} kp_k. \tag{8.8}$$

Again taking the first order difference of (8.8) and taking $r(m) = q_m/q_{m+1}$, we obtain for $m = 0,1,\ldots,N-2$

$$\frac{(m+b)q_m^2 - (m+1+b)q_m q_{m+1}}{p_m} - \frac{(m+1+b)q_{m+1}^2 - (m+2+b)q_{m+1}q_{m+2}}{p_{m+1}} = mp_m.$$

Dividing both sides of the above equation by q_{m+1} and writing $r(m) = q_m/q_{m+1}$ we obtain for $m = 0,1,\ldots,N-2$

$$\frac{(m+b)r^2(m) - (m+1+b)r(m)}{r(m)-1} - \frac{(m+1+b)r(m+1) - (m+2+b)}{r(m+1)-1} = m(r(m)-1)$$

Putting $h(m) = 1/(r(m)-1)$, we get from the above equation for $m = 0,1,\ldots,N-2$

$$h(m+1) = 2 + h(m) - b/h(m). \tag{8.9}$$

Assume $N = \infty$ and $h(0) = b/2$, then the solution of (8.9) is $q_j = \left(\frac{b}{b+2}\right)^j$, $j = 0,1,\ldots$ □

Example 7. (Beta-Binomial distribution) Let $X_1, X_2, \ldots$ be a sequence of independent and identically random variables having Beta-Binomial distribution, $BB(N, \beta)$ with pmf

$$p(j) = P(X = j) = \binom{\beta + N - j - 1}{N - j}\binom{\beta + N}{N}^{-1}, \quad j = 0.1,, \ldots, N$$

and

$$\begin{aligned} \bar{P}(k-1) = q_k &= \sum_{j=k}^{N} p_j \\ &= \sum_{j=k}^{N} \binom{\beta + N - j - 1}{N - j}\binom{\beta + N}{N}^{-1} \\ &= \binom{\beta + N - k}{N - k}\binom{\beta + N}{N}^{-1}. \end{aligned}$$

Here

$$\begin{aligned} R_{w,1}(x_2) &= \sum_{1 \le x_1 \le x_2} \frac{p(x_1)}{\bar{P}(x_1 - 1)} \\ &= \sum_{1 \le x_1 \le x_2} \frac{\binom{\beta + N - x_1 - 1}{N - x_1}}{\binom{\beta + N - x_1}{N - x_1}} \\ &= \sum_{1 \le x_1 \le x_2} \frac{\beta}{N + n - x_1}. \end{aligned}$$

Thus, for $j = k, \ldots, N$

$$\begin{aligned} p_{w,2}(j) &= R_{w,1}(j)p(j) \\ &= \sum_{0 \le x_1 \le j} \frac{\beta}{N + n - x_1}\binom{\beta + N - j - 1}{N - j}\binom{\beta + N}{N}^{-1} \end{aligned}$$

and

$$\begin{aligned} E(X_{U_w(n)} | X_{U_w(n-2)} = m) &= \frac{1}{q_m} \sum_{r=m}^{N} \frac{p_r}{q_r} \sum_{k=r}^{N} k p_k \\ &= \binom{\beta + N - m}{N - m}^{-1} \sum_{r=m}^{N} \frac{\beta}{\beta + N - r} \sum_{k=r}^{N} k \binom{\beta + N - k - 1}{N - k}. \end{aligned}$$

Now

$$\begin{aligned} \sum_{k=r}^{N} k \binom{\beta + N - k - 1}{N - k} &= \sum_{i=0}^{N-r} \binom{\beta + i - 1}{i} \\ &= N \sum_{i=0}^{N-r} \binom{\beta + i - 1}{i} - \sum_{i=0}^{N-r} i \binom{\beta + i - 1}{i} \\ &= N \binom{\beta + N - r}{n - r} - \binom{\beta + N - r}{N - r - 1} \end{aligned}$$

$$
\begin{aligned}
&= N\binom{\beta+N-r}{N-r} - \frac{(N-r)\beta}{\beta+1}\binom{\beta+N-r}{N-r} \\
&= \frac{N+r\beta}{\beta+1}\binom{\beta+N-r}{N-r}.
\end{aligned}
$$

Thus

$$
E(X_{U_w(n)}|X_{U_w(n-2)} = m) = \frac{\beta}{\beta+1}\binom{\beta+n-m}{n-m}^{-1}\sum_{r=m}^{N}\frac{N+r\beta}{\beta+N-r}\binom{\beta+N-r}{N-r}.
$$

Now

$$
\begin{aligned}
\sum_{r=m}^{N}\frac{N+r\beta}{\beta+N-r}\binom{\beta+N-r}{N-r} &= \sum_{r=m}^{N}\frac{N+r\beta}{\beta}\binom{\beta+N-r-1}{N-r} \\
&= \frac{N}{\beta}\binom{\beta+N-m}{N-m} - \sum_{r=m}^{N} r\binom{\beta+N-r-1}{N-r}.
\end{aligned}
$$

It can easily be shown that

$$
\begin{aligned}
\sum_{r=m}^{N} r\binom{\beta+N-r-1}{N-r} &= \sum_{t=0}^{N-m}(N-t)\binom{\beta+t-1}{t} \\
&= N\binom{\beta+N-m}{N-m} - \sum_{t=0}^{N-m} t\binom{\beta+t-1}{t} \\
&= N\binom{\beta+N-m}{N-m} - \frac{(N-m)\beta}{\beta+1}\binom{\beta+N-m}{N-m} \\
&= \frac{N+m\beta}{\beta+1}\binom{\beta+N-m}{N-m}.
\end{aligned}
$$

Thus

$$
\begin{aligned}
E(X_{U_w(n)}|X_{U_w(n-2)} = m) &= \left(\frac{\beta}{\beta+1}\right)\left(\frac{N}{\beta} + \frac{N+m\beta}{\beta+1}\right) \\
&= \left(\frac{\beta}{\beta+1}\right)^2 m + \left(\frac{2\beta+1}{(\beta+1)^2}\right)N.
\end{aligned}
$$

Wesolowski and Ahsanullah (2001) proved that if $E(X_{U_w(n)}|X_{U(n-2)} = m) = am + b$, where $a < 1$, then the distribution of the X's is Beta-Binomial.

Ahsanullah and Hijab (2006) proved the following characterization theorem.

Theorem 14. *Let $\{X_n,\ n \geq 1\}$ be a sequence of independent and identically distributed random variables with common distribution function F. Suppose X_n is concentrated on $0, 1, \ldots \infty$. Then $X_j \in GE(p)$ iff $X_{U_w(n)} - X_{U_w(n-1)} \stackrel{d}{=} W$, $n > 1$ where W is independent of $X_{U_w(n-1)}$ and has the same distribution as X's.*

For a generalization of Theorem 14, see Ahsanullah (2008).

References

[1] Ahsanullah, M. *Record Values - Theory and Applications*. University Press of America: Lanham, MD, 2006.

[2] Ahsanullah, M. Some characteristic properties of weak records of geometric distribution. *J. Stat. Theo. and Appl.* 2008, *7*, 1, 81–92.

[3] Ahsanullah, M.; Hijab, O. Some characterizations of geometric distribution by weak records. In Ahsanullah, M., Raqab, M. Z. (Eds.) *Recent Developments in Ordered Random Variables*, Nova Publishers: New York, 2006, pp. 187–195.

[4] Ahsanullah, M.; Holland, B. On the Use of Record Values to Estimate the Location and Scale Parameters of the Generalized Extreme Value Distribution. *Sankhya (Series A)* 1994, *56*, 480–499.

[5] Ahsanullah, M.; Nevzorov, V. B. *Ordered Random Variables*. Nova Science Publishers: New York, 2001.

[6] Aliev, F. A. Characterizations of Discrete Distributions through Weak Records. *J. Appl. Satist. Sci.* 1998, *8*, 13–16.

[7] Aliev, F. A.; Ahsanullah, M. Characterizations of discrete distributions through regression of record values. *Pakistan J. Statistics* 2002, *18*, 3, 415–421.

[8] Bairamov, I. G.; Ahsanullah, M. Distributional Relations Between Order Statistics and the Sample itself and Characterizations of Exponential Distribution. *J. Appl. Stat. Sci.* 2000, *10*, 1, 1–16.

[9] Galambos, J.; Seneta, E. Record Times. *Proc. Amer. Math. Soc.* 1975, *50*, 383–387.

[10] Srivastava, R. C. Two Characterizations of the Geometric Distribution by Record Values.*Sankhya (Series B)* 1978, *40*, 276–278.

[11] Stepanov, A. V. A Characterization Theorem for Weak Records. *Theory Prob. Appl.* 1994, *38*, 762–764 (English Translation).

[12] Vervaat, W. Limit Theorems for Records from Discrete Distributions. *Stoch. Proc. Their Appl.* 1973, *1*, 317–334.

[13] Wesolowski, J.; Ahsanullah, M. Linearity of Regression for Non-adjacent Weak Records. *Statistica Sinica* 2001, *11*, 30–52.

[14] Williams, D. Rényi's Record Problem and Engel's Series. *Bulletin of the London Mathematical Society* 1973, *5*, 235–237.

In: Records and Branching Processes
Editor: M. Ahsanullah, G. P. Yanev, pp. 27-35
ISBN 978-1-60456-526-3

Chapter 2

CHARACTERIZATIONS OF DISTRIBUTIONS VIA BIVARIATE REGRESSION ON DIFFERENCES OF RECORDS

***Ilham Akhundov*[1] *and Valery B. Nevzorov*[2*]**
[1] Statistics and Actuarial Science, University of Waterloo,
200 University Avenue West, Waterloo, Ontario, Canada N2L 3G1
[2] Department of Mathematics and Mechanics
Saint Petersburg State University, Starii Petergoff
Universitetskii Prospekt 28, Saint Petersburg, 198504 Russia

Abstract

We consider record values $X(n), n > 0$ generated from a sequence of independent random variables $X_1, X_2, X_3, \ldots$ with a common continuous distribution function. In this chapter we characterize the distributions such that for $0 < x < y < \infty$ we have $E(X(3) - X(2)|X(1) = x, X(4) = y) = (y - x)/3$.

Key words and phrases. Characterizations, record values, regression.

1 Introduction

Let $X, X_1, X_2, \ldots$ be independent random variables (rv's) having a common cumulative distribution function (cdf) $F(x)$. Define the so-called upper (strong) record times $L(n)$ and the corresponding record values $X(n)$ as $L(1) = 1$, $X(1) = X_1$, and for $n > 1$,

$$L(n) = \min\{j : X_j > X(n-1)\} \quad \text{and} \quad X(n) = X_{L(n)} = \max\{X_1, \ldots, X_{L(n)}\}.$$

The theory of records is reviewed in details in the monographs by Arnold, Balakrishnan and Nagaraja (1998), and Nevzorov (2000). Among different results obtained for records, characterizations of distributions based on record values have been discussed extensively in the literature. Many interesting characterizations have been established by regression

*E-mail address: iakhundo@math.uwaterloo.ca; decanat@math.spbu.ru

relations of one record value on one or two other record values (cf. [3-7] and Sections 4.2.2 and 4.3.2 in [1]).

Bairamov, Ahsanullah and Pakes (2005) characterize the exponential family of distributions by the bivariate regression function

$$E(\varphi(X(n))|X(n-1)=x,X(n+1)=y), \quad n=2,3,\ldots$$

for some functions φ. As a particular case of their result, the following holds

$$E(X(2)|X(1)=x,X(3)=y)=\frac{x+y}{2},$$

for all $l_F<x<y<r_F$, where $l_F=\inf\{t:F(t)>0\}$ and $r_F=\sup\{t:F(t)<1\}$, characterizes the exponential distributions with cdf

$$F(x)=1-\exp\{-c(x-a)\} \quad x>a, \quad c>0, \quad -\infty<a<\infty. \tag{1.1}$$

The above characterization property can be interpreted as follows: the conditional expectations of the spacings $X(2)-X(1)$ and $X(3)-X(2)$ given fixed values for $X(1)$ and $X(3)$ are equal , that is

$$\begin{aligned} E(X(2)-X(1)|X(1)=x,X(3)=y) &= E(X(3)-X(2)|X(1)=x,X(3)=y) \\ &= \frac{y-x}{2} \end{aligned}$$

if and only if $F(x)$ has the form of (1.1). Consider now the four record values $X(1)<X(2)<X(3)<X(4)$. In this paper we first show that for the standard exponential distribution for $0<x<y<\infty$

$$\begin{aligned} &E(X(2)-X(1)|X(1)=x,X(4)=y) \\ &= E(X(3)-X(2)|X(1)=x,X(4)=y) \\ &= E(X(4)-X(3)|X(1)=x,X(4)=y) \\ &= \frac{y-x}{3} \end{aligned} \tag{1.2}$$

Then we verify that separately taken from (1.2) the equality for $0<x<y<\infty$

$$E(X(3)-X(2)|X(1)=x,X(4)=y)=\frac{y-x}{3} \tag{1.3}$$

holds for at least one more family of distributions. Finally, we find all distributions characterized by (1.3).

2 Results and Proofs

First, let us prove that (1.3) is valid for the standard exponential distribution. It is known that the vector $(X(1),X(2),X(3),X(4))$ has the same distribution as the vector $(\xi_1,\xi_1+$

$\xi_2, \xi_1+\xi_2+\xi_3, \xi_1+\xi_2+\xi_3+\xi_4)$, where ξ_1, ξ_2, ξ_3 and ξ_4 are independent rv's having the standard exponential distribution. Then

$$\begin{aligned}
&E(X(3)-X(2)|X(1)=x, X(4)=y)\\
&= E(\xi_3|\xi_1=x, \xi_1+\xi_2+\xi_3+\xi_4=y)\\
&= E(\xi_3|\xi_1=x, \xi_2+\xi_3+\xi_4=y-x)\\
&= E(\xi_3|\xi_2+\xi_3+\xi_4=y-x)\\
&= \frac{E(\xi_2+\xi_3+\xi_4|\xi_2+\xi_3+\xi_4=y-x)}{3}\\
&= \frac{y-x}{3}.
\end{aligned}$$

Similarly, for the standard exponential distribution, setting $k=2,3,\ldots,n$ we obtain

$$E(X(k)-X(k-1)|X(1)=x, X(n)=y) = \frac{y-x}{n-1}. \tag{2.1}$$

Moreover, it is easy to see that any linear transform of the standard exponential distribution preserves (2.1). It follows that (1.3) and (2.1) are valid for the cdf (1.1).
It turns out that (1.3) is true for at least one more family of distributions. To show this, consider the distribution with cdf $F(x) = 1-\exp\{-x^{1/2}\}, x>0$. It is evident that in this case

$$(X(1), X(2), X(3), X(4)) \stackrel{d}{=} (\xi_1^2, (\xi_1+\xi_2)^2, (\xi_1+\xi_2+\xi_3)^2, (\xi_1+\xi_2+\xi_3+\xi_4)^2),$$

where ξ_1, ξ_2, ξ_3, and ξ_4 are, as above, independent rv's having the standard exponential distribution. Then

$$\begin{aligned}
&E(X(3)-X(2)|X(1)=x, X(4)=y)\\
&= E((\xi_1+\xi_2+\xi_3)^2-(\xi_1+\xi_2)^2|\xi_1^2=x, (\xi_1+\xi_2+\xi_3+\xi_4)^2=y)\\
&= E(\xi_3(2\xi_1+2\xi_2+\xi_3)|\xi_1=x^{1/2}, \xi_2+\xi_3+\xi_4=y^{1/2}-x^{1/2})\\
&= E(\xi_3(2x^{1/2}+2\xi_2+\xi_3)|\xi_1=x^{1/2}, \xi_2+\xi_3+\xi_4=y^{1/2}-x^{1/2})\\
&= E(\xi_3(2x^{1/2}+2\xi_2+\xi_3)|\xi_2+\xi_3+\xi_4=y^{1/2}-x^{1/2})\\
&= 2x^{1/2}E(\xi_3|\xi_2+\xi_3+\xi_4=y^{1/2}-x^{1/2})+E(2\xi_2\xi_3+\xi_3^2|\xi_2+\xi_3+\xi_4=y^{1/2}-x^{1/2}).
\end{aligned}$$

It is evident that

$$E(\xi_3|\xi_2+\xi_3+\xi_4=y^{1/2}-x^{1/2}) = \frac{y^{1/2}-x^{1/2}}{3}.$$

Also, one can see that the random variables $\nu_1 = 2\xi_2\xi_3+\xi_3^2$, $\nu_2 = 2\xi_3\xi_1+\xi_1^2$ and $\nu_3 = 2\xi_1\xi_2+\xi_2^2$ are symmetrically dependent, and $\nu_1+\nu_2+\nu_3 = (\xi_2+\xi_3+\xi_4)^2$. Hence

$$\begin{aligned}
&E(2\xi_2\xi_3+\xi_3^2|\xi_2+\xi_3+\xi_4\\
&= y^{1/2}-x^{1/2})\\
&= \frac{E((\xi_2+\xi_3+\xi_4)^2|\xi_2+\xi_3+\xi_4=y^{1/2}-x^{1/2})}{3}\\
&= \frac{(y^{1/2}-x^{1/2})^2}{3},
\end{aligned}$$

and thus we have

$$
\begin{aligned}
&E(X(3)-X(2)|X(1)=x,X(4)=y)\\
&= \frac{2x^{1/2}(y^{1/2}-x^{1/2})+(y^{1/2}-x^{1/2})^2}{3}\\
&= \frac{y-x}{3}.
\end{aligned}
$$

Therefore, (1.3) is valid at least for two different families of distributions. Below we will find all distributions characterized by (1.3).

2.1 Main result

Suppose, in addition, that the rv's $X_1, X_2, \ldots$ have a density function $f(x)$. Then the following result is valid.

Theorem 1. *The regression equality*

$$E(X(3)-X(2)|X(1)=x,X(4)=y) = \alpha x+\beta y+\gamma \; (l_F < x < y < r_F) \qquad (2.2)$$

holds if and only if $\beta = -\alpha = 1/3, \gamma = 0, l_F > -\infty, r_F = \infty$, *and* $F(x)$ *(up to location and scale parameters) is given by either*

$$F(x) = 1-\exp\{-x\} \quad \text{if} \quad x \geq 0; \quad F(x) = 0 \quad \text{if} \quad x < 0,$$

or

$$F(x) = 1-\exp\{-x^{1/2}+c^{1/2}\} \quad \text{if} \quad x > c; \quad F(x) = 0 \quad \text{if} \quad x \leq c, \qquad (2.3)$$

where $c \geq 0$ *is a shape parameter.*

Remark. One can see that along with the exponential family of distributions, relation (2.2) holds for all distributions with cdf

$$F(x) = 1-\exp\Big\{-\Big(\frac{x-a}{b}\Big)^{1/2}+c^{1/2}\Big\}, \qquad x > cb+a,$$

where $c \geq 0, b > 0$ and $-\infty < a < \infty$. Notice also that the above cdf $F(x) = 1-\exp\{-x^{1/2}\}$, $x > 0$, belongs to the family of distributions (2.3) with $c = 0$.

2.2 Proof of the theorem.

Denote $R(x) = f(x)/(1-F(x))$. The joint density $f_n(x_1, x_2, \ldots, x_n)$ of the record values $X(1), X(2), \ldots, X(n)$ has the form (see, for example, (15.9) in Nevzorov (2000))

$$f_n(x_1, x_2, \ldots, x_n) = \begin{cases} R(x_1)\cdots R(x_n)(1-F(x_n)) & x_1 < x_2 < \cdots < x_n, \\ 0 & \text{otherwise.} \end{cases} \qquad (2.4)$$

Let $f_{2,3|1,4}(u, v|x, y)$ denote the conditional density function of $X(2)$ and $X(3)$ given that $X(1) = x$ and $X(4) = y$, where $x < u < v < y$. The following result is valid.

Lemma 1. *For any $x < u < v < y$,*

$$f_{2,3|1,4}(u,v|x,y) = \frac{2R(u)R(v)}{[\ln(1-F(x)) - \ln(1-F(y))]^2} \tag{2.5}$$

Proof of Lemma 1. The joint probability density function $f_{1,4}(x,y)$ of the records $X(1)$ and $X(4)$ can easily be found from (2.4) as

$$f_{1,4}(x,y) = \frac{R(x)R(y)(1-F(y))\,(\ln(1-F(x)) - ln(1-F(y)))^2}{2}, \quad x < y. \tag{2.6}$$

Then (2.5) immediately follows from the relation

$$f_{2,3|1,4}(u,v|x,y) = \frac{f_4(x,u,v,y)}{f_{1,4}(x,y)},$$

taking into account (2.4) (for $n = 4$), and (2.6). Lemma 1 is proved, which enables us to get a simple proof (we will omit it) of the next lemma. □

Lemma 2. *For any $x < y$,*

$$E(X(3) - X(2)|X(1) = x, X(4) = y)$$
$$= \frac{2\int_x^y tR(t)[\ln(1-F(x) - 2\ln(1-F(t)) + \ln(1-F(y))]dt}{[\ln(1-F(x)) - \ln(1-F(y))]^2}.$$

Let us continue the proof of the theorem. By Lemma 1, the condition (2.2) can be rewritten as

$$2\int_x^y tR(t)[\ln(1-F(x) - 2\ln(1-F(t)) + \ln(1-F(y))]dt$$
$$= (\alpha x + \beta y + \gamma)(\ln(1-F(x)) - \ln(1-F(y)))^2.$$

If we let x tend to y in (2.2), then it is evident that the left-hand side of (2.2) tends to zero, and for any y one obtains that $(\alpha+\beta)y + \gamma = 0$. Therefore $\gamma = 0$ and $\alpha + \beta = 0$, that is $\alpha = -\beta$. Since $0 < X(3) - X(2) < X(4) - X(1)$ one can conclude that $0 < \beta < 1$. Hence, we must study the relation

$$\beta(y-x)(\ln(1-F(x)) - \ln(1-F(y)))^2$$
$$= 2\int_x^y tR(t)[\ln(1-F(x) - 2\ln(1-F(t)) + \ln(1-F(y))]dt, \tag{2.7}$$

with $0 < \beta < 1$. For simplicity, denote $H(x) = -\ln(1-F(x))$ and $\delta = 2/\beta$, where $\delta > 2$. Note that $H'(x) = R(x)$. By differentiating (2.7) with respect to x and y, and simplifying we obtain

$$\frac{(H(y) - H(x))(H'(x) + H'(y))}{y-x} = \frac{\delta-2}{2}H'(x)H'(y), \qquad x < y. \tag{2.8}$$

Letting x to tend to y again, we obtain

$$2(H'(y))^2 = \frac{\delta-2}{2}(H'(y))^2. \tag{2.9}$$

Since there are points y such that $H'(y) > 0$, it follows from (2.9) that $\delta = 6$ (hence $\beta = 1/3$), and thus (2.8) can be rewritten as

$$H(y) - H(x) = \frac{2H'(x)H'(y)(y-x)}{H'(x)+H'(y)}. \tag{2.10}$$

Consider the possible cases for this problem. Suppose that $l_F = -\infty$ and pick any y, $l_F < y < r_F$, such that $f(y) > 0$ and let x tend to $-\infty$ in (2.10). Since $H(-\infty) = 0, 0 < H(y) < \infty, H'(y) > 0$, the only possibility for (2.10) to hold is the case, when $f(x) \to 0$, and hence $H'(x) \to 0$, as $x \to -\infty$. But in this situation we obtain from (2.10) that for any y, $H(y) = -2\lim_{x\to-\infty} xf(x)$, that is $H(y)$ does not depend on y. This contradiction implies $l_F > -\infty$. Thus, we need to investigate the equation

$$E(X(3) - X(2)|X(1) = x, X(4) = y) = \frac{y-x}{3} \qquad l_F < x < y < r_F, \tag{2.11}$$

for the case $l_F > -\infty$. It is evident that linear transforms of the initial rv's X's preserve (2.11). It follows that we can (without loss of generality) set $l_F = 0$. Letting x to tend to l_F $(= 0)$ we are to consider the following four cases.

Case 1. Let $\lim_{x\to 0} f(x) = 0$ and thus $H'(x) = f(x)/(1-F(x)) \to 0$, as $x \to 0$. Since $H(0) = 0$, it follows from (2.10) that $H(y) = 0$ for all y, where $f(y) > 0$. This is impossible.

Case 2. Let $\lim_{x\to 0} f(x) = \infty$. It follows that $\lim_{x\to 0} H'(x) = \infty$ and hence

$$H(y) = 2yH'(y), \qquad 0 < y < r_F. \tag{2.12}$$

It follows from (2.12) (under condition $H(0) = 0$) that $H(y) = cy^{1/2}, c > 0$, and then $F(y) = 1 - \exp\{-cy^{1/2}\}, \quad y \geq 0$.

Case 3. Let $\lim_{x\to 0} f(x) = d$, $0 < d < \infty$. Then one obtains from (2.10) the following equation

$$H(y) = \frac{2dyH'(y)}{d+H'(y)}, \qquad 0 < y < r_F.$$

Considering a scale transform (taking $dX_1, dX_2, \ldots$ instead of the initial X's) we can suppose without restriction that $d = 1$ and consider the equation

$$H(y) = \frac{2yH'(y)}{1+H'(y)},$$

which is equivalent to

$$H'(y) = \frac{H(y)}{2y - H(y)}, \qquad 0 < y < r_F, \tag{2.13}$$

where $0 \leq H(y) < 2y$ since $0 \leq H'(y) < \infty$. Introduce the new function $G(y)$ such that $H(y) = yG(y)$. Together with $H(y)$, the function $G(y)$ is continuous on the interval $(0, r_F)$. We have from (2.13) that

$$yG'(y) = \frac{G^2(y) - G(y)}{2 - G(y)}.$$

Note that if $G(y) = 1$ for some y, then $H(y) = y$ and $F(y) = 1 - \exp\{-y\}$ for the same y. Since $G(y)$ is a continuous function, we have $1 < G(y) < 2$. For all y, such that $G(y) \neq 1$, we have the differential equation

$$\frac{1}{y} = \frac{G'(y)(2 - G(y))}{G(y)(G(y) - 1)},$$

the solution of which can be expressed as

$$Cy = \frac{1 - G(y)}{G^2(y)}. \tag{2.14}$$

Here $C > 0$ if $0 < G(y) < 1$, and $C < 0$ if $1 < G(y) < 2$. Recalling that $G(y)$ is continuous, we see that if there is at least one interval, where $0 < G(y) < 1$, and one, where $1 < G(y) < 2$, then there must be a point y^* such that $G(y^*) = 1$ and $G(y^* - 0) < 1$ or $G(y^* - 0) > 1$. Letting y to tend to y^* in (2.14) we obtain $Cy^* = 0$ and $y^* = 0$. This contradicts to our assumption that we have both types of intervals where $0 < G(y) < 1$ and where $1 < G(y) < 2$. The same method allows us to conclude that if $G(y) = 1$ for some point $y > 0$, then $G(y) = 1$ for any $y > 0$. It follows that only the following three options need to be considered:

(i) $G(y) \equiv 1$ for all $y > 0$;

(ii) $0 < G(y) < 1$, for all $y > 0$;

(iii) $1 < G(y) < 2$, for all $y > 0$.

Assume (i). Then $H(y) = y, y > 0$, and hence $F(x) = 1 - \exp\{-x\}, x > 0$. Assume (ii). Then there is a positive constant C such that

$$Cy = \frac{1 - G(y)}{G^2(y)}, \qquad y > 0.$$

The solution of this equation has the form

$$G(y) = -\frac{1}{2Cy} \pm \left(\frac{1}{4C^2y^2} + \frac{1}{Cy}\right)^{1/2} \qquad y > 0.$$

Since $G(y)$ must be positive we obtain

$$G(y) = \left(\frac{1}{4C^2y^2} + \frac{1}{Cy}\right)^{1/2} - \frac{1}{2Cy}, \qquad y > 0,$$

and hence

$$H(y) = \left(\frac{1}{4C^2} + \frac{y}{C}\right)^{1/2} - \frac{1}{2C}, \qquad y > 0.$$

Then

$$F(y) = 1 - \exp\left\{-\left(\frac{1}{4C^2} + \frac{y}{C}\right)^{1/2} + \frac{1}{2C}\right\}, \qquad y > 0. \tag{2.15}$$

Analogously, assuming (iii) we have that the solution has the form

$$G(y) = -\frac{1}{2Cy} \pm \left(\frac{1}{4C^2y^2} + \frac{1}{Cy}\right)^{1/2},$$

where C now is a negative constant. Since $|Cy| = (G(y)-1)/G^2(y)$ and $1 < G(y) < 2$, it is easy to see that $|Cy| \leq 1/4$. Then $-1/(2Cy) \geq 2$, and therefore we have

$$G(y) = -\frac{1}{2Cy} - \left(\frac{1}{4C^2y^2} + \frac{1}{Cy}\right)^{1/2}, \qquad 0 < y < \frac{|C|}{4},$$

$$H(y) = \frac{1}{2C} - \left(\frac{1}{4C^2} + \frac{y}{C}\right)^{1/2},$$

and

$$F(y) = 1 - \exp\left\{\left(\frac{1}{4C^2} + \frac{y}{C}\right)^{1/2} - \frac{1}{2C}\right\}, \qquad 0 < y < \frac{1}{4|C|}. \tag{2.16}$$

Evidently, the function (2.16) can not be a distribution function and thus (iii) does not give us new solutions.

Case 4. The last situation that needs to be considered, is the case when $f(0)$ does not exist. However, it follows from (2.10) that the limit

$$\lim_{x\downarrow 0} \frac{f(x)}{f(x)+H'(y)} = \frac{H(y)}{yH'(y)} \tag{2.17}$$

exists for any y. It is not difficult to see that (2.17) requires the existence of $\lim_{x\downarrow 0} f(x)$, which contradicts the assumption in this case.

Therefore, we have found all possible types of cdf's for which equation (2.2) holds. Taking into account the possible linear transforms of the initial random variables we can conclude that the families of distributions (2.15) and (2.3) coincide. A simple verification shows that the families of distributions included in the theorem do satisfy (2.2). The proof is complete.

Acknowledgment

This work was partially supported by RFBR (grants 06-01-00763 and 06-01-00179) and the Natural Sciences and Engineering Research Council of Canada

References

[1] Ahsanullah, M.; Nevzorov, V. B. Characterizations of distributions by regressional properties of records. *J. Appl. Statist. Science* 2004, *13*, 1:33–39.

[2] Arnold, B. C.; Balakrishnan, N.; Nagaraja, H. N. *Records*. Wiley: New York, 1998.

[3] Bairamov, I.; Ahsanullah, M.; Pakes, A. A characterization of continuous distributions via regression on pairs of record values. *Austr. N. Z. J. Statist*. 2005, *47*, 543–547.

[4] Gupta R. C.; Ahsanullah, M. Some characterization results based on the conditional expectation of a function of non-adjacent order statistics (record values). *Ann. Inst. Statist. Math.* 721–732, 2004.

[5] Nagaraja, H. N. On a characterization based on record values. *Australian Journal of Statistics* 1977, *19*, 70–73.

[6] Nagaraja, H. N. On the expected values of record values.*Australian Journal of Statistics* 1978, *20*, 176–182.

[7] Nevzorov, V. B. Records: *Mathematical Theory.* Translations of Mathematical Monographs 194, American Math. Soc.: New York, 2000.

In: Records and Branching Processes
Editor: M. Ahsanullah, G. P. Yanev, pp. 37-62
ISBN 978-1-60456-526-3

Chapter 3

REGENERATIVE BRANCHING PROCESSES

Kosto V. Mitov*[1] *and Nikolay M. Yanev*[2]
[1] Aviation Faculty, National Military University
5856 Dolna Mitropolia, Bulgaria
[2] Department of Probability and Statistics
Institute of Mathematics and Informatics
Bulgarian Academy of Sciences, 8
Acad. G. Bonchev Street, 1113 Sofia, Bulgaria

Abstract

In this review we present some results on branching models having regeneration at zero as their characteristic feature, obtained in the last thirty years. Regenerative phenomena also arise in the evolution of branching processes with homogeneous migration. However, the analytical methods of studying these processes do not use the regenerative property explicitly. Here we present some results obtained using a recently developed probabilistic technique based on the renewal and regenerative properties of processes with migration.

Key words and phrases. Branching processes, Regeneration, Immigration, Emigration, Migration, Limit theorems.

1 Introduction

It is well known that the classical branching processes eventually become extinct (the population dies out) with probability one in both critical and subcritical cases. For this reason the limiting behavior of the processes is investigated conditionally on non-extinction. In the seventies Foster and Pakes introduced models of branching processes with state-dependent immigration. Processes of this type allow immigration of new particles only in state zero, i.e., the population regenerates when it becomes extinct. Thus, zero is not an absorbing state anymore and becomes a reflecting barrier. The process evolution consists of a sequence of cycles which are independent and stochastically equivalent. The immigration in state zero might be viewed as an alternative to the conditioning on non-extinction.

*E-mail address: kmitov@af-acad.bg; yanev@math.bas.bg

Bienaymé-Galton-Watson (BGW) branching processes with state-dependent immigration were studied by Foster [4] and Pakes [21], [22], and [23]. Later the following generalizations were proposed: Markov branching processes with continuous time (Yamazato [35]); Bellman-Harris branching processes (Mitov and Yanev [14], [15] and Slavchova and Yanev [27], [28]); branching diffusion processes (Li [10]).

In this review we present some results on the above branching models, having regeneration at zero as their characteristic feature, obtained in the last thirty years. Regenerative phenomena also arise in the evolution of branching processes with homogeneous migration. However, the analytical methods of studying these processes do not use the regenerative property explicitly. Here we present some results obtained using a recently developed probabilistic technique based on the renewal and regenerative properties of processes with migration.

The article is organized as follows. In Section 2 we give a general definition that includes all models of regenerative branching processes considered by now. Section 3 presents the main results for the classical Foster and Pakes model. Section 4 discusses the first generalization of this model to the case of continuous time Markov branching processes. Section 5 presents an extension to the class of age-dependent branching processes. Section 6 contains the definition and some general limit theorems for alternating regenerative processes. An application to the Bellman-Harris branching processes with state-dependent immigration is given in the same section. Section 7 includes a Foster type limit theorem for continuous time and state branching processes with state dependent immigration. Section 8 is devoted to another application of the basic results from Section 6 to BGW processes with homogeneous migration. In each of these sections after stating the definition of a particular model, we compare it with the general definition in Section 2. The last Section 9 presents a generalization of the classical Foster-Pakes process, namely BGW branching processes with time non-homogeneous state-dependent immigration.

2 General Definition

Let us have on a probability space $(\Omega, \mathcal{A}, \mathbf{P})$ three independent sets of random variables as follows.

(i) The set $Z = \{\{Z_{kj}(t), t \geq 0\},\ k, j = 1, 2, \ldots.\}$ of independent, identically distributed (i.i.d.) branching processes with probability generating function (p.g.f.) $F(t;s) = \mathbf{E}\{s^{Z_{kj}(t)}\}$, $|s| \leq 1$, and $Z_{kj}(0) = 1$ almost surely (a.s.), i.e. $F(0;s) = s$.

(ii) The set $I = \{I_k,\ k = 1, 2, \ldots\}$ of i.i.d. integer valued positive random variables (r.v.) with p.g.f. $g(s) = \mathbf{E}\{s^{I_k}\}$, $|s| \leq 1$. (the number of immigrants in the state zero.)

(iii) The set $X = \{X_k,\ k = 1, 2, \ldots\}$ of i.i.d., positive r.v. with cumulative distribution function (c.d.f.) $A(t) = \mathbf{P}\{X_k \leq t\}$, $t \geq 0$ (stay at zero or waiting period).

Let us define the sequence $\{\{Z_k(t), t \geq 0\},\ k = 1, 2, \ldots\}$ of i.i.d. branching processes starting with a positive random number of ancestors I_k at time $t = 0$, as follows

$$Z_k(t) = \sum_{j=1}^{I_k} Z_{kj}(t), \quad t \geq 0.$$

By the independence of the sets Z and I it follows that

$$\Phi(t,s) := \mathbf{E}\{s^{Z_k(t)}\} = g(F(t,s)), \quad |s| \leq 1.$$

Denote by T_k the life period (time to extinction) of the process $Z_k(t)$, i.e. T_k is a r.v. such that

$$Z_k(0) = I_k > 0,\ Z_k(t) > 0 \text{ for } t \in [0, T_k), \text{ and } Z_k(T_k) = 0.$$

It is clear that the r.v. $\{T_k,\ k = 1,2,\ldots\}$ are i.i.d. and the equivalence of the events $\{Z_k(t) = 0\}$ and $\{T_k \leq t\}$ yields

$$B(t) = \mathbf{P}\{T_k \leq t\} = \Phi(t,0) = g(F(t,0)).$$

Consider the sequence $Y_k = X_k + T_k,\ \ k = 1,2,\ldots,$ of i.i.d. positive r.v. and denote their c.d.f. by

$$D(t) = \mathbf{P}\{Y_k \leq t\} = \int_0^t A(t-u)dB(u) = (A * B)(t).$$

Define the renewal epochs

$$S_0 = 0,\ \ S_{n+1} = S_n + Y_{n+1},\ \ n = 0,1,2,\ldots$$

and the ordinary renewal process

$$N(t) = \max\{n : S_n \leq t\}.$$

The *alternating renewal sequence*

$$\{(S_n,\ S'_{n+1}),\ n = 0,1,2,\ldots\}, \quad \text{where } S'_{n+1} = S_n + X_{n+1}, \quad n = 0,1,2,\ldots$$

allows us to define the process $\sigma(t)$, connected with the moment t,

$$\sigma(t) = t - S'_{N(t)+1} = t - (S_{N(t)} + X_{N(t)+1}).$$

If $\sigma(t) \geq 0$ then it can be interpreted as the "spent lifetime." If $\sigma(t) < 0$ then $-\sigma(t) > 0$ is the rest of the "waiting period" (see Figure 1). (For more details see Mitov and Yanev [16], [17].)

Now we are in a position to define the branching process with state-dependent immigration $Z(t)$ for $t \geq 0$ as follows

$$Z(t) = Z_{N(t)+1}(\sigma(t))\mathbb{I}_{\{\sigma(t)\geq 0\}}. \tag{2.1}$$

Here and later $\mathbb{I}_A$ denotes the indicator of the event A.

The process $\{Z(t), t \geq 0\}$ has the following interpretation (see Figure 2). It starts at time $t = 0$ by $Z(0) = 0$ and stays at state 0 for a random period $S'_1 = X_1$. At time S'_1, there is an immigration into the process of I_1 (of age 0) particles ($I_1 > 0, a.s.$). The immigrants initiate a branching process $Z_1(t - S'_1)$ and $Z(t)$ coincides with this process up to time $S_1 = X_1 + T_1$ when $Z_1(t - X_1)$ hits the state zero. During the "waiting periods" $[S_{n-1}, S'_n)$, the process $Z(t)$ equals zero, while in the "life periods" $[S'_n, S_n)$ the process $Z(t)$ coincides with the corresponding branching process initiated by I_n immigrants in the state 0.

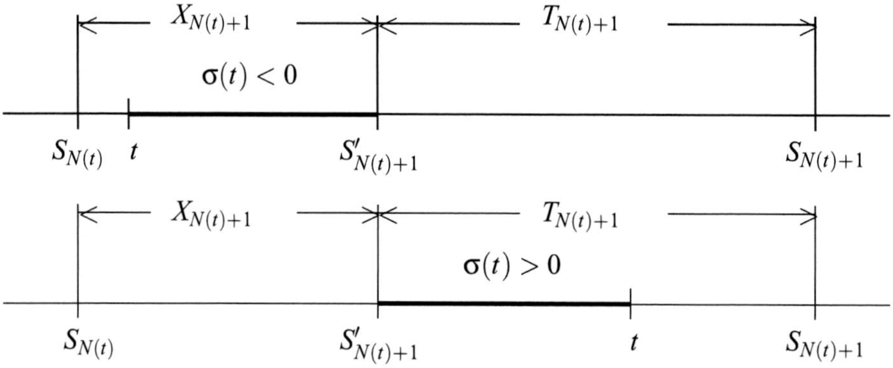

Figure 1: The spent time $\sigma(t)$.

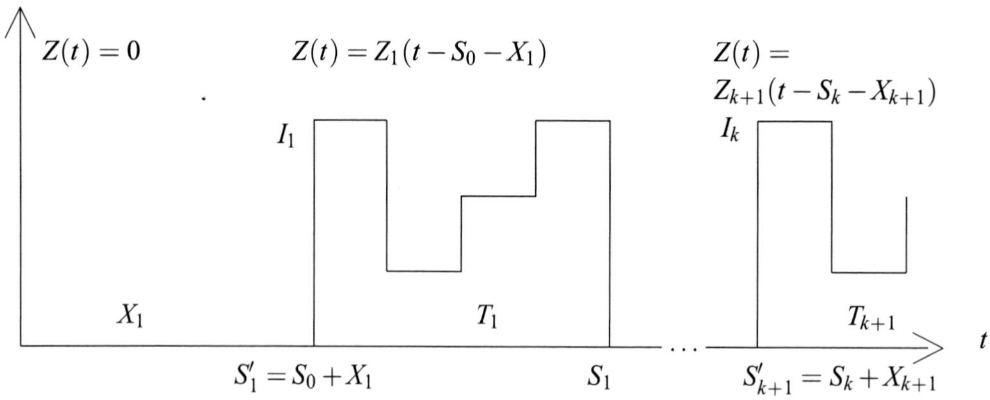

Figure 2: Regenerative branching process

Remarks. (i) In the definition above the parameter $t \geq 0$ can be continuous or discrete. (ii) The conditions $Z(0) = 0$ and $Z(t) = 0$, for $t \in [0, X_1]$ are not restrictive. They simplify the notation, making the corresponding renewal and regenerative processes without delay.

In the following sections we will discuss some particular cases of the general definition above. In each individual case we will point out how it can be obtained for the general definition.

3 BGW Processes with State-dependent Immigration

First, in this section, we discuss the classical Foster-Pakes model. Let us have (see [21]) two probability generating functions $f(s) = \sum_{n=0}^{\infty} f_n s^n$, and $h(s) = \sum_{n=0}^{\infty} h_n s^n$, where $|s| \leq 1$. Define the homogeneous Markov chain $\{Z(t), t = 0, 1, 2, \ldots\}$ with one step transition probabilities

$$p_{ij} = \mathbf{P}\{Z(t+1) = j | Z(t) = i\}, \quad i, j = 0, 1, 2, \ldots$$

determined by the following equation for the probability generating functions

$$\sum_{j=0}^{\infty} p_{ij} s^j = (1 - \delta_{i0})[f(s)]^i + \delta_{i0} h(s),$$

where δ_{ij} is the Kronecker delta. This Markov chain is irreducible and aperiodic provided that

$$0 < f_0, f_0 + f_1, h_0 < 1, f_1 > 0. \tag{3.1}$$

Denote the critical parameter $m = f'(1)$ and the mean number of immigrants by $m_I = h'(1)$.

The above definition can be obtained from the general one given in Section 2 imposing the following conditions:

(i) The time parameter is discrete $t = 0, 1, 2, \ldots$

(ii) $Z_{kj}(t), t = 0, 1, 2, \ldots$ are BGW processes, each having a single ancestor.

It can be seen that the distribution of the immigrants $\{h_j, j = 0, 1, 2, \ldots\}$ and the assumption $h_0 < 1$ lead to geometric distribution with parameter $1 - h_0$ of the stay at zero X_k, i.e. $\mathbf{P}\{X_k = j\} = (1 - h_0) h_0^{j-1}, j = 1, 2, \ldots$. At the end of the zero cycle, $I_k > 0$ new particles immigrate where $g_j = \mathbf{P}\{I_k = j\} = h_j / (1 - h_0), j = 1, 2, \ldots$.

Therefore, we have to assume in the general definition that also

(iii) The random variables X_k have geometric distribution with $\mathbf{P}\{X_k = j\} = (1 - p) p^{j-1}$, $j = 1, 2, \ldots$, then setting $h(s) = p + (1 - p) g(s)$, (i.e. $h_0 = p, h_j = (1 - p) g_j, j = 1, 2, \ldots$).

Remark. The process $\{Z(t), t = 0, 1, 2, \ldots\}$, defined by Pakes, begins with a certain number of particles $Z(0) = i \geq 0$. On the other hand, in the general definition the process starts with $Z(0) = 0$. Since the Markov chain is irreducible the states belong to one class, and this difference is not essential.

Pakes [21]-[23] studies all three cases: subcritical, critical and supercritical. He investigates the ergodicity of the Markov chain $\{Z(t), t = 0, 1, 2 \ldots\}$ and proves limit theorems that we present next.

Theorem 1 (Pakes, [21]). *Suppose that $m < 1$ (subcritical process) and (3.1) holds. The process has a stationary distribution*

$$\lim_{t \to \infty} \mathbf{P}\{Z(t) = j\} = \pi_j, \quad j = 0, 1, 2, \ldots,$$

if and only if $\sum_{j=2}^{\infty} h_j \log j < \infty$. The stationary distribution satisfies for $|s| \leq 1$

$$\sum_{j=0}^{\infty} \pi_j s^j = 1 - \pi_0 \sum_{t=0}^{\infty} (1 - h(f_t(s)), \qquad \pi_0 = \left(1 + \sum_{t=0}^{\infty} (1 - h(f_t(0))\right)^{-1}.$$

Theorem 2 (Pakes, [21]). *Let $m > 1$ (supercritical process) and (3.1) holds.*

(i) *There exists a sequence* $C_t, t = 0, 1, 2, \ldots$ *of positive numbers such that* $C_{t+1}/C_t \to m$ *as* $t \to \infty$, *and*

$$\frac{Z(t)}{C_t} \xrightarrow{a.s.} Z, \text{ as } t \to \infty.$$

The Laplace-Stieltjes transform of the limit Z is given by

$$\mathbf{E}\{e^{-\lambda Z}\} = [\phi(\lambda)]^i - \sum_{t=0}^{\infty}[1 - h(\phi(\lambda/m^t))](F(t;0))^i, \quad 0 \le \lambda < \infty,$$

where i is the initial number of particles and $\phi(\lambda)$ *satisfies the equation* $\phi(m\lambda) = f(\phi(\lambda))$. *The numbers* $C_t \sim \text{const}.m^t$ *if and only if* $\sum_{j=2}^{\infty} f_j j \log j < \infty$.

(ii) *The Markov chain* $\{Z(t), t = 0, 1, 2, \ldots\}$ *is geometrically ergodic, i.e.,*

$$\mathbf{P}\{Z(t) = j | Z(0) = i\} = q_i d_j R^{-t} + o(R^{-t}), \quad t \to \infty, \quad (i, j = 0, 1, 2, \ldots)$$

where the constants $R, (1 < R < (f'(q))^{-1})$, q_i *and* d_j *are explicitly determined in terms of the offspring and immigration distributions and* $q > 1$ *is the probability of extinction of the processes* $\{Z_{kj}(t), t = 0, 1, 2, \ldots\}$.

Remark. (i) Due to the fact that in the supercritical case the distribution of interarrival times is degenerate (it has a mass $1 - q$ at infinity), the renewal process $\{S_n, n = 0, 1, 2, \ldots\}$ is transient, i.e. with probability one there exists a last renewal epoch $M > 0$. Therefore, the process coincides with $\{Z_{k_M}(t - M), t = 0, 1, 2, \ldots\}$ after time M. This, partially, explains the fact that the normalizing sequence $\{C_t\}$ is the same as in a simple BGW process.

The results in the critical case concern the existence of an invariant measure and the asymptotics of the nth step transition probabilities.

Theorem 3 (Pakes, [21]). *Let* $m = 1$ (*critical case*) *and* (3.1) *holds. If* $m_I < \infty$, *and* $f''(1) = 2b < \infty$, *then the process* $\{Z(t), t = 0, 1, 2 \ldots\}$ *is null-recurrent and*

$$\frac{m_I}{b}\frac{\log t}{t}\sum_{k=0}^{t} \mathbf{P}\{Z(k) = j | Z(0) = i\} \to \mu_j, \quad t \to \infty, \quad (i, j = 0, 1, 2, \ldots),$$

where $\{\mu_j, j = 0, 1, 2, \ldots\}$ *is the invariant measure. If, additionally,* $\sum_{j=2}^{\infty} h_j j \log j < \infty$, *then* $\sum_{j=0}^{l} \mu_j \sim \log l$ *as* $l \to \infty$.

One might expect that the limiting distribution in the critical case is independent on the particular choice of offspring and immigration distributions. The following theorem confirms this.

Theorem 4 (Foster, [4]). *Assume* $m = 1$ (*critical process*) *and* (3.1). *If* $0 < f''(1) < \infty$, $m_I < \infty$, *and* $f_j < 1$ *for all* $j = 0, 1, 2, \ldots$, *then*

$$\lim_{t \to \infty} \mathbf{P}\left\{\frac{\log Z(t)}{\log t} \le x\right\} = x, \quad x \in (0, 1).$$

At the end of his paper, Foster gives, without proof, a limiting distribution when the number of immigrants has an infinite mathematical expectation. This case is studied in details by Pakes [22] who proves a limit theorem in case when the mean number of immigrants is infinite and the offspring variance is finite. He also comments on the case of infinite offspring variance.

Theorem 5 (Pakes [22]). *Suppose* $m = 1$, $2b = f''(1) < \infty$, *and* $1 - h(s) = (1-s)^{\theta} L(1/(1-s))$, $(0 < \theta < 1)$, *where* $L(\cdot)$ *is slowly varying at infinity. If* $0 < \theta \leq 1/2$, *assume in addition that the sequence* $h(0)$, $h(f_{n-1}(0)) - h(f_{n-2}(0))$, $n = 2, 3, \ldots$ *is decreasing. Then* $Z(t)/(bt)$ *converges in law to a random variable with Laplace-Stieltjes transform*

$$\widehat{M}_1(\lambda) = 1 - \frac{\sin \pi\theta}{\pi} \int_0^1 (\lambda^{-1} + x)^{-\theta} (1-x)^{\theta-1} dx.$$

An important question for a Markov chain is whether it is recurrent or transient. Let us point out that, under the assumption that the Markov chain is aperiodic and irreducible, this is the same as the question whether the corresponding renewal process $\{S_n\}_{n=0}^{\infty}$ is recurrent (persistent) or transient. Pakes answers this question in his paper [21] (see Theorem 1, Theorem 3, and Theorem 6 in [21]). An extension of his results is proved by Nakagawa and Sato [20] as given below.

Theorem 6 (Nakagawa and Sato [20]). *Let the Markov chain* $\{Z(t), t = 0, 1, 2, \ldots\}$ *be irreducible and aperiodic* (*i.e. conditions* (3.1) *hold*).

(i) *Necessary and sufficient conditions for the chain to be positive recurrent* (*i.e., to have a stationary distribution*) *are*

$$m \leq 1 \quad and \quad \int_0^1 \frac{1 - h(s)}{f(s) - s} ds < \infty.$$

(ii) *If* $m = 1$, $m_I < \infty$, *and* $f''(1) < \infty$, *then* $\{Z(t)\}$ *is null recurrent.*

(iii) *If* $m > 1$, *then* $\{Z(t)\}$ *is transient.*

Kulkarni and Pakes [8] investigate further these processes. They study the limiting behavior of the total progeny of the BGW processes with state-dependent immigration. Define the total progeny of the process $\{Z(t), t = 0, 1, 2, \ldots\}$ up to time (generation) t as

$$W(t) = \sum_{k=0}^{t} Z(k), \quad t = 0, 1, 2, \ldots.$$

The regenerative property of $\{Z(t)\}$ and the properties of the renewal process $\{S_n, n = 0, 1, 2, \ldots\}$ play a major role in the investigation together with the functional equations for the probability generating functions. As one may expect, in the subcritical case the renewal process $\{S_n, n = 0, 1, 2, \ldots\}$ is positive recurrent (see Theorem 6 above). This leads to a strong law of large numbers and asymptotic normality of $\{W(t)\}$.

Theorem 7 (Kulkarni and Pakes [8]). *Assume* (3.1). *If the process is subcritical* $(m < 1)$ *and* $m_I < \infty$, *then*

$$\frac{W(t)}{t} \overset{a.s.}{\rightarrow} \frac{m_I \pi_0}{1-m}, \quad t \to \infty.$$

If additionally, $f''(1) < \infty$ and $Var I_k < \infty$, then

$$\frac{W(t) - m_I N(t)/(1-m)}{\sqrt{t}} \xrightarrow{d} N(0, \pi_0 \nu^2),$$

and

$$\frac{W(t) - m_I \pi_0 t/(1-m)}{\sqrt{t}} \xrightarrow{d} N(0, \pi_0 \omega^2),$$

where the constants ν and ω are explicitly calculated.

In the critical case the authors prove that $N(t)\log t/t \xrightarrow{P} 1/\gamma$, as $t \to \infty$, where the positive constant γ is explicitly determined by the moments of offspring and immigration. This leads to the following limit theorem.

Theorem 8 (Kulkarni and Pakes [8]). *Assume* (3.1). *If the process is critical* $(m = 1)$, $0 < 2b = f''(1) < \infty$, *and* $m_I < \infty$, *then*

$$\left(\frac{\log t}{t}\right)^2 W(t) \xrightarrow{d} \frac{b}{2} W_{1/2}, \quad t \to \infty,$$

where $W_{1/2}$ is a standard stable random variable with index $1/2$.

Next result holds in the supercritical case.

Theorem 9 (Kulkarni and Pakes [8]). *Assume* (3.1). *If the process is supercritical* $(1 < m < \infty)$, *then*

$$\frac{W(t)}{C_t} \xrightarrow{a.s.} \frac{m}{m-1} Z, \quad t \to \infty,$$

where the random variable Z is as in Theorem 2.

It is not surprising that the normalizing sequence $\{C_t\}$ is the same as for $Z(t)$. This is due to the fact that the underlying renewal process is transient as we mentioned before.

4 Continuous Time Markov Processes

Yamazato [35] proposes a generalization of the Foster-Pakes model for continuous time Markov branching processes. Assume that $\{p_i, i = 0, 1, 2, \ldots\}$ are such that $p_1 < 0$, $p_i \geq 0$, $i \neq 1$, and $\sum_{i=0}^{\infty} p_i = 0$; and $\{q_j, j = 0, 1, 2, \ldots\}$ are such that $q_0 < 0$, $q_j \geq 0$, $j = 1, 2, \ldots$, and $\sum_{j=0}^{\infty} q_j = 0$.

Yamazato's definition is in terms of the transition kernel of the continuous time Markov chain.

Definition (Yamazato [35]). Let $\{Z(t), t \geq 0\}$ be continuous time Markov chain on non-negative integers whose transition kernel is

$$A = (a_{ij}), \text{ where } a_{ij} = \begin{cases} i p_{j-i+1}, & \text{if } j \geq i-1 \text{ and } i \geq 1, \\ q_j, & \text{if } i = 0, \\ 0, & \text{otherwise}. \end{cases}$$

This definition can be viewed as a particular case of the general definition in Section 2 by making the following assumptions.

(i) Assume that $t \in [0, \infty)$ is continuous.

(ii) The processes $Z_{kj}(t), t \geq 0$, are continuous time Markov branching processes with infinitesimal p.g.f. $f(s) = \sum_{i=0}^{\infty} p_i s^i$, $|s| < 1$, $p_1 < 0$, $\sum_{i=0}^{\infty} p_i = 0$. The critical parameter is $\alpha = f'(1)$.

(iii) The random variables X_k are exponentially distributed with parameter $0 < -q_0 < 1$ and the distribution of the immigrants at zero I_k is $\mathbf{P}\{I_k = l\} = \dfrac{q_l}{1+q_0}$, $l = 1, 2, 3, \ldots$. Denote by $g(s) = \sum_{k=0}^{\infty} q_k s^k$, $|s| < 1$ the infinitesimal p.g.f. of the immigration and let $\beta = g'(1)$.

Yamazato investigates in details the question whether the Markov chain is transient, null-recurrent or positive recurrent. He also proves limit theorems in the critical and super-critical cases.

Theorem 10 (Yamazato [35]). *The process $Z(t)$ is recurrent iff $\alpha \leq 0$.*

(i) *If additionally, $\int_0^1 \frac{f(s)}{g(s)} ds > -\infty$, then $Z(t)$ is positive recurrent.*

(ii) *If the integral in (i) diverges the process is null-recurrent.*

Theorem 11 (Yamazato [35]). *If $0 < \alpha < \infty$ and $\beta < \infty$, then $Z(t)e^{-\alpha t}$, $t \geq 0$ converges almost surely to some positive random variable Z. If additionally, $f''(1) < \infty$ and $g''(1) < \infty$ then as $t \to \infty$*

$$\mathbf{E}\{Z(t)|Z(0) = i\} \sim C_1 e^{\alpha t} \quad \textit{and} \quad \mathbf{E}\{Z^2(t)|Z(0) = i\} \sim C_2 e^{2\alpha t},$$

where the constants $C_i, i = 1, 2$ are explicitly given.

Theorem 12 (Yamazato [35]). *Let $\alpha = 0$, $\beta < \infty$, and $f''(1) < \infty$. Then*

$$\lim_{t \to \infty} \mathbf{P}\left\{ \frac{\log Z(t)}{\log t} \leq x \right\} = x, \quad x \in (0, 1).$$

If additionally, $g''(1) < \infty$, then

$$\mathbf{E}\{Z(t)|Z(0) = i\} = i + \beta \int_0^t P(Z(u) = 0|Z(0) = i) du$$

and

$$\mathbf{E}\{Z^2(t)|Z(0) = i\} = i^2 + \left(f''(1)(\beta+1) + g''(1)\right) \int_0^t P(Z(u) = 0|Z(0) = i) du.$$

5 Age-Dependent Branching Processes

In a series of papers, Mitov and Yanev ([14], [15]) and Slavtchova and Yanev ([27], [28]) extend the Foster-Pakes model to age-dependent branching processes. Although these processes are not Markovian, the regenerative property observed in the Markov case allows the general definition given in Section 2. In this section we assume the conditions stated there. Assume also that the processes $Z_{kj}(t), t \geq 0$ are age-dependent branching processes governed by the offspring p.g.f. $f(s)$ and let $G(t)$, $(G(0+) = 0)$ be the c.d.f. of the life-length of a particle. It is well known that the p.g.f. $F(t;s)$ of the process $Z_{kj}(t)$ is the unique p.g.f. solution of the integral equation

$$F(t;s) = s(1 - G(t)) + \int_0^t f(F(t-u;s)dG(u), \quad F(0;s) = s.$$

Using the regenerative property of the process, Mitov and Yanev [14] obtain the following linear integral equation

$$R(t;s) = \int_0^t R(t-u;s)dD(u) + \int_0^t Q(t-u;s)dA(u),$$

where $R(t;s) = 1 - \mathbf{E}\{s^{Z(t)}\}$ and $Q(t;s) = 1 - F(t;s)$, $s \in [0,1)$, $t \geq 0$. This equation is the main tool in the study of the age-dependent branching processes with state-dependent immigration in the papers [14] and [15], where the authors consider the critical case. The main results are collected in the following theorem. Further on we will use the incomplete beta function $\mathrm{B}_x(a,b) = \int_0^x u^{a-1}(1-u)^{b-1}du$, for $x \in [0,1]$ and $a, b > 0$.

Theorem 13 (Mitov and Yanev [14, 15]). *Assume $m = 1$ (critical case), $0 < 2b = f''(1) < \infty$, $r = \int_0^\infty x dG(x) < \infty$, and $t^2[1 - G(t)] \to 0$ as $t \to \infty$.*

(i) *If $\mathbf{E}\{X_i\} < \infty$ and $m_I = g'(1) < \infty$, then*

$$\lim_{t\to\infty} \mathbf{P}\left\{\frac{\log Z(t)}{\log t} \leq x\right\} = x, \quad 0 < x < 1.$$

(ii) *If $\mathbf{E}\{X_i\} < \infty$ and $g(s) = 1 - (1-s)^\theta L_1((1-s)^{-1})$, $\theta \in (1/2, 1)$, then*

$$\lim_{t\to\infty} \mathbf{P}\left\{\frac{Z(t)}{t} \leq x\right\} = M_1(x), x \geq 0, \tag{5.1}$$

where $M_1(x)$ is a distribution function with Laplace transform

$$\widehat{M_1}(\lambda) = 1 - \mathrm{B}_{\lambda/(\lambda+r/b)}(\theta, 1-\theta)/\mathrm{B}(\theta, 1-\theta), \quad \lambda > 0.$$

(iii) *If $\mathbf{E}\{X_i\} = \infty$, $g(s) = 1 - (1-s)^\theta L_1((1-s)^{-1})$ for $\theta \in (1/2, 1)$, and $1 - A(t) = L_2(t)t^{-\alpha}$ with $\alpha > \theta$ then (5.1) holds.*

(iv) *If $\mathbf{E}\{X_i\} = \infty$, $g(s) = 1 - (1-s)^\theta L_1((1-s)^{-1})$ for $\theta \in (1/2, 1)$, and $1 - A(t) = L_2(t)t^{-\alpha}$ with $\alpha \leq \theta$ then*

$$\lim_{t\to\infty} \mathbf{P}\left\{\frac{Z(t)}{t} \leq x | Z(t) > 0\right\} = M_2(x), \quad x \geq 0,$$

where $M_2(x)$ is a distribution function with Laplace transform

$$\widehat{M_2}(\lambda) = 1 - [\lambda/(\lambda + r/b)]^{\theta-\alpha}\mathrm{B}_{\lambda/(\lambda+r/b)}(\alpha, 1-\theta)/\mathrm{B}(\alpha, 1-\theta), \quad \tau > 0.$$

(v) *If $\mathbf{E}\{X_i\} = \infty$, $1 - A(t) = L_2(t)t^{-\alpha}$ and $m_I = g'(1) < \infty$, then*

$$\lim_{t\to\infty} \mathbf{P}\left\{\frac{\log Z(t)}{\log t} \leq x | Z(t) > 0\right\} = x, \quad 0 < x < 1.$$

Note that the assertions (i) and (ii) are direct generalizations of Theorem 4 and Theorem 5 respectively. Furthermore, it is clear from Theorem 13 that the limiting distribution essentially depends on the relation between the length of the stay at zero (zero cycle) and the length of the intervals when it evolves (cycle of evolution). The limiting distribution can be conditional or unconditional depending on the lengths of these two cycles. If the zero cycle is 'much longer' than the cycle of evolution then the limit distribution is conditional (cases (iv) and (v)); whereas if the evolution cycle is 'longer' or of 'the same infinite length' as the zero cycle, then the limiting distribution is unconditional. On the other hand the length of T_i depends on the criticality and on the immigration distribution.

The above theorem deals with the situation when the processes $Z_{kj}(t)$ are critical with *finite variance.* In this case the following result is central for the proof.

Proposition (Goldstein [5]). *If* $m = 1$, $0 < 2b = f''(1) < \infty$, $r = \int_0^\infty t dG(t) < \infty$, *and* $t^2[1 - G(t)] \to 0$ *then uniformly in* $0 \leq s < 1$ *as* $t \to \infty$ *we have*

$$1 - F(t;s) = \frac{1-s}{1 + \frac{bt}{r}(1-s)}(1 + o(1)).$$

In particular, if $s = 0$ we have $\mathbf{P}\{Z_{kj}(t) > 0\} \sim r/(bt)$. The last relation together with the immigration distribution determine the tail behavior of the c.d.f. $\mathbf{P}\{T_i \leq t\} = B(t)$. On the other hand, the Proposition does not hold when the offspring *variance is infinite* when the tail behavior of T_i and the asymptotic of the processes $\{Z_{ij}(t)\}$ change significantly. Notice that the critical Markov branching processes with finite and infinite variance have quite different limiting behavior too. The critical BGW processes without immigration and infinite offspring variance was investigated by Slack [26] and Zolotarev [43]. Pakes [22] discussed the limiting behavior of the processes with state-dependent immigration and infinite offspring variance. We will consider the infinite variance case later in Section 6.

Now we present results for the non-critical cases.

Theorem 14 (Slavtchova and Yanev [28]). *Assume* $m < 1$ *(subcritical case),* $r = \int_0^\infty x dG(x) < \infty$, $\mathbf{E}\{X_i\} < \infty$, $m_I = g'(1) < \infty$, *and the c.d.f.* $G(\cdot)$ *and* $A(\cdot)$ *are non-lattice. Then there exists limiting distribution*

$$\lim_{t\to\infty} \mathbf{P}\{Z(t) = k\} = \pi_k, k = 0, 1, 2, \ldots,$$

where

$$\phi(s) = \sum_{k=0}^{\infty} \phi_k s^k = 1 - \frac{Q(s)}{\mathbf{E}\{I_i\}}, \quad Q(s) = \int_0^\infty Q(t;s)dt, \ |s| \leq 1.$$

Denote by μ the Malthusian parameter corresponding to the pair $(f(s), G(t))$, i.e. μ is the unique solution of the equation

$$m \int_0^\infty e^{-\mu u} dG(u) = 1.$$

Theorem 15 (Slavtchova and Yanev [28]). *Assume* $m > 1$ *(supercritical case),* $2b = f''(1) < \infty$, $m_I = g'(1) < \infty$ *and* $g''(1) < \infty$. *Then the process* $W(t) = Z(t)e^{-\mu t}$ *converges in* *mean square and almost surely to W whose Laplace-Stieltjes transform satisfies the equation $\psi(\lambda) = \int_0^\infty \psi(\lambda e^{-\mu u})dD(u) + \int_0^\infty f(\phi(\lambda e^{-\mu u}))dA(u) - f(q)$ where $\phi(\lambda)$ is the unique solution of the equation $\phi(\lambda) = \int_0^\infty h(\phi(\lambda e^{-\mu u}))dG(u)$ in the class of Laplace-Stieltjes transforms of nonnegative proper random variables.*

Slavtchova and Yanev [28] study also the asymptotic behavior of the first and second moments.

6 Alternating Regenerative Processes

Mitov and Yanev [17] study general processes dropping the assumption in Section 2 definition that the processes $Z_k(t)$ in the evolution cycles are branching processes. This eliminates the possibility of using the probability generating functions analysis. The authors obtain some new limit theorems for *alternating regenerative process*. (For a complete review on renewal and regenerative processes see e.g. Kalashnikov [7] or Sigman and Wolf [25].)

Assume that the alternating regenerative processes satisfy the following conditions.

Condition 1. *The distribution functions $A(t) = \mathbf{P}\{X_k \le t\}$ and $B(t) = \mathbf{P}\{T_k \le t\}$ are non-lattice, $A(0) = B(0) = 0$ and there exists*

$$\lim_{t\to\infty} \frac{1 - A(t)}{1 - B(t)} = c, \quad 0 \le c \le \infty. \tag{6.1}$$

Condition 2.

$$m_A \equiv \mathbf{E}\{X_k\} < \infty \tag{6.2}$$

or

$$\mathbf{E}\{X_k\} = \infty, \qquad 1 - A(t) \sim t^{-\alpha} L_A(t), \qquad \alpha \in (1/2, 1], \tag{6.3}$$

and for every $h > 0$ fixed, $A(t) - A(t-h) = O(1/t),\ t > 0$.

Condition 3.

$$m_B \equiv \mathbf{E}\{T_k\} < \infty, \tag{6.4}$$

or

$$\mathbf{E}\{T_k\} = \infty, \qquad 1 - B(t) \sim t^{-\beta} L_B(t), \qquad \beta \in (1/2, 1]. \tag{6.5}$$

The functions $L_A(\cdot)$ and $L_B(\cdot)$ are slowly varying at infinity.

Condition 4. *The asymptotic behavior of the processes $\{Z_k(t), t \ge 0\}$ on the 'life cycle' is given by*

$$\lim_{t\to\infty} \mathbf{P}\left\{ \frac{Z_k(t)}{M(t)} \le x | T_k > t \right\} = H(x), \tag{6.6}$$

where $H(x)$ is a proper d.f. on $\mathbf{R}^+$ and $M(t)$ is a positive, non-decreasing function, regularly varying at infinity with parameter $\gamma \ge 0$, i.e., $M(t) = t^\gamma L_M(t)$, where $L_M(t)$ is slowly varying at infinity.

Denote $m_A(t) = \int_0^t (1 - A(x))dx$ and $m_B(t) = \int_0^t (1 - B(x))dx$. Under the above conditions the following fundamental theorem holds.

Theorem 16 (Mitov and Yanev [17]). *Assume conditions* (6.1) *and* (6.6).
1) *Let* (6.5) *holds with* $1/2 < \beta < 1$.
(i) *If additionally,* (6.2) *or* (6.3) *is fulfilled and* $0 \le c < \infty$, *then*

$$\lim_{t\to\infty} \mathbf{P}\left\{\frac{Z(t)}{M(t)} \le x\right\} = \frac{c}{1+c} + \frac{G_1(x)}{1+c},$$

where

$$G_1(x) = \frac{\sin \pi\beta}{\pi} \int_0^1 H(xu^{-\gamma})(1-u)^{\beta-1}u^{-\beta}du$$

is a d.f. on $\mathbf{R}^+$;
(ii) *If* (6.3) *is fulfilled and* $c = \infty$, *then*

$$\lim_{t\to\infty} \mathbf{P}\left\{\frac{Z(t)}{M(t)} \le x | Z(t) > 0\right\} = G_2(x),$$

where

$$G_2(x) = \frac{1}{\mathrm{B}(1-\beta,\alpha)} \int_0^1 H(xu^{-\gamma})(1-u)^{\alpha-1}u^{-\beta}du$$

is a d.f. on $\mathbf{R}^+$ *and* $\mathrm{B}(\cdot,\cdot)$ *is the standard beta function.*
2) *Let* (6.5) *holds with* $\beta = 1$, $H(0) = 0$ *and* $\gamma > 0$.
(iii) *If additionally,* (6.2) *or* (6.3) *is fulfilled and* $0 \le c < \infty$, *then for* $0 < x < 1$

$$\lim_{t\to\infty} \mathbf{P}\left\{\frac{m_B(M^{-1}(Z(t)))}{m_B(t)} \le x\right\} = \frac{c}{1+c} + \frac{x}{1+c},$$

where $M^{-1}(\cdot)$ *is the inverse function of* $M(\cdot)$;
(iv) *If* (6.3) *is fulfilled and* $c = \infty$, *then for* $0 < x < 1$

$$\lim_{t\to\infty} \mathbf{P}\left\{\frac{m_B(M^{-1}(Z(t)))}{m_B(t)} \le x | Z(t) > 0\right\} = x.$$

3) *Let* (6.4) *and* (6.3) *hold. Then*

$$\lim_{t\to\infty} \mathbf{P}\{Z(t) \le x | Z(t) > 0\} = \frac{1}{m_B} \int_0^\infty \mathbf{P}\{Z_1(u) \le x, T_1 > u\}du.$$

The proof of Theorem 16 is based on the properties of alternating renewal processes studied in [16]. A discrete-time version is given in [11].

Remarks. (i) If one assumes (6.2) and (6.4) only, then the process $Z(t)$ with renewal epochs $\{S_n\}$ is positive recurrent and without any other assumptions

$$\lim_{t\to\infty} \mathbf{P}\{Z(t) \le x\} = \frac{1}{m_A + m_B} \int_0^\infty \mathbf{P}\{Z(u) \le x, X_1 + T_1 > u\}du.$$

by the classical Feller's regenerative theorem (cf. [25], Theorem 2.1).
(ii) If $M(t) \sim t$ and $m_B(t) \sim \log t$ then

$$\frac{m_B(M^{-1}(Z(t)))}{m_B(t)} \sim \frac{\log Z(t)}{\log t}, \quad (t \to \infty).$$

(iii) It is not difficult to verify the following implications:
1) $[(6.3) \wedge (6.5) \wedge (\alpha > \beta)] \vee [(6.2) \wedge (6.5))] \Rightarrow c = 0$;
2) $[(6.3) \wedge (6.5) \wedge (\alpha < \beta)] \vee [(6.3) \wedge (6.4))] \Rightarrow c = \infty$;
3) $[(6.3) \wedge (6.5) \wedge (\alpha = \beta)] \vee [(6.2) \wedge (6.4))] \Rightarrow 0 \le c \le \infty$.

Applying Theorem 16 it is not difficult to obtain some well-known results for renewal processes with heavy tails. Indeed, let $X_k \equiv 0$ *a.s.* Then by (6.1) it follows that $c = 0$. Let $Z_k(t) = t - S_{k-1}, S_{k-1} \le t < S_k$, and let $M(t) \equiv t$. Then by (6.6) one obtains that $H(x) = 0$ for $x < 1$ and $H(x) = 1$ for $x \ge 1$. Note that by the definition (2.1) we have $Z(t) \equiv \sigma(t)$, i.e., the classical spent lifetime. Now from Theorem 16(i)

$$\lim_{t\to\infty} \mathbf{P}\left\{\frac{\sigma(t)}{t} \le x\right\} = \frac{\sin \pi\beta}{\pi} \int_0^1 (1-u)^{\beta-1} u^{-\beta} du, \quad x \ge 0,$$

which is the result proved by Dynkin [2] and Lamperti [9]. Similarly by Theorem 16(iii) leads to the theorem of Erickson [3]

$$\lim_{t\to\infty} \mathbf{P}\left\{\frac{m_B(\sigma(t))}{m_B(t)} \le x\right\} = x, \quad 0 \le x \le 1.$$

It is worth mentioning that the assumption (6.6) inherits the limiting properties of critical branching processes. This explains the regular variation of the normalizing function $M(\cdot)$. Therefore, some generalization of the theorem is possible in this direction. Another generalization might be a possible dependence between the lengths of the 'zero cycles' and the 'life cycles'. On the other hand, in its current form the theorem is general enough to serve as a tool in further investigations of branching processes with regenerative properties. Here we will discuss one application of Theorem 16 to critical branching processes with infinite variance and state-dependent immigration.

The limiting behavior of a critical branching process with infinite offspring variance differs significantly from that in the finite variance case. The first results for infinite offspring variance were due to Zolotarev [43] and Slack [26]. The behavior of the age-dependent branching processes depends also on the asymptotic of the ratio $\dfrac{n(1-G(n))}{1-f_n(0)}$ as $n \to \infty$ (see (6.9) below.) This asymptotic was studied in details by Vatutin (see e.g. [29-33]). With these results at hand and Theorem 16 stated above, we are in a position to investigate the critical branching processes with a state-dependent immigration and infinite offspring variance. Assume that the offspring p.g.f. has the form

$$f(s) = s + (1-s)^{1+\delta} L\left(\frac{1}{1-s}\right), \quad \delta \in (0,1], \tag{6.7}$$

where $L(t)$ is slowly varying at infinity. The condition (6.7) defines $Z_{kj}(t)$ as critical Bellman-Harris processes with possibly infinite variance. For the life-length c.d.f. we assume

$$G(0) = 0, \quad G(t) \text{ is non-lattice and } r = \int_0^\infty x dG(x) < \infty, \tag{6.8}$$

and

$$\lim_{n\to\infty} \frac{n(1-G(n))}{1-f_n(0)} = d, \quad 0 \le d \le \infty, \tag{6.9}$$

where, as usual, $f_n(s)$ denotes the nth fold iteration of $f(s)$: $f_0(s) = s$, $f_{n+1}(s) = f(f_n(s))$. Note that in the Markovian case the limit of the above ratio is uniquely determined.

For the immigration at zero we assume one of the following conditions

$$m_I = \mathbf{E}\{I_k\} < \infty, \tag{6.10}$$

or

$$\mathbf{E}\{I_k\} = \infty \quad \text{and} \quad g(s) = 1 - (1-s)^{\theta} L_B(\frac{1}{1-s}), \ \theta \in (0,1], \tag{6.11}$$

where $L_B(t)$ is slowly varying at infinity. Finally, for the waiting periods we assume one of the conditions (6.2) or (6.3).

In a series of papers Vatutin ([29],[30],[31],[32], [33]) studies critical age-dependent branching processes with infinite offspring variance. He obtains different types of limit distributions depending on the limit in (6.9). In the framework of regenerative processes, Vatutin's results give the conditional limiting behavior of the process in separate life cycles and provide (6.6). This construction is explored in [18] to prove limit theorems for critical age-dependent branching processes with infinite variance and state-dependent immigration. Since the limit distributions depend on the combination of the values of the parameters c, δ, d, θ, and α the set of possible limit theorems is quite large. Here we present a limit theorem from [18] that corresponds to the case $0 < d < \infty$. For the complete results we refer the reader to [18].

Let us introduce some notation. Following Vatutin [29], denote by $\rho(s)$ the unique positive solution of the equation

$$\rho^{1+\delta}(s) - r\rho(s) - (1-s)d\delta = 0, \qquad s \in [0,1],$$

where $\delta \in (0,1]$ is from (6.7), $0 < r < \infty$ is from (6.8), and $0 < d < \infty$ is from (6.9). We will also need the following Laplace-Stieltjes transform

$$\widehat{D}_{\rho,\delta,\theta}(\lambda) = 1 - \lambda^{\theta}\Big(1 + \frac{(\rho(0)\lambda)^{\delta}}{r}\Big)^{-\theta/\delta}, \ \lambda > 0. \tag{6.12}$$

Recall that $Q(t;s) = 1 - \mathbf{E}\{s^{Z_{ij}(t)}\}$ $(s \in [0,1])$ and $Q(t) = \mathbf{P}\{Z_{ij}(t) > 0\}$.

Theorem 17 (Mitov and Yanev [18]). *Assume* (6.7)–(6.9) *with* $0 < d < \infty$ *and* (6.1). A) *Suppose additionally* (6.10), (6.3) *and either* $0 < \delta < 1$ *or* $\delta = 1$ *with* $\mathbf{E}\{T_1\} < \infty$ *holds. Then*

$$\lim_{t\to\infty} \mathbf{P}\{Z(t) \le x | Z(t) > 0\} = \frac{1}{m_B}\int_0^{\infty} \mathbf{P}\{Z_1(u) \le x, T_1 > u\} du. \tag{6.13}$$

B) *Suppose additionally* (6.11) *and denote* $\beta = \theta/\delta$.

1) *If* $1/2 < \beta < 1$ *and either* (6.2) *or* (6.3) *with* $\beta < \alpha \le 1$ *holds, then*

$$\lim_{t\to\infty} \mathbf{P}\{Z(t)Q(t) \le x\} = \frac{\sin \pi\beta}{\pi}\int_0^1 D_{\rho,\delta,\theta}(xu^{-1/\delta})(1-u)^{\beta-1}u^{-\beta} du = G_1(x) \tag{6.14}$$

where the d.f. $D_{\rho,\delta,\theta}(x)$ *has Laplace transform* $\widehat{D}_{\rho,\delta,\theta}(\lambda)$ *given in (6.12).*

2) *If* (6.3) *with* $1/2 < \alpha < \beta < 1$ *is fulfilled, then*

$$\lim_{t\to\infty} \mathbf{P}\{Z(t)Q(t) \le x | Z(t) > 0\} = \frac{1}{\mathbf{B}(1-\beta,\alpha)}\int_0^1 D_{\rho,\delta,\theta}(xu^{-1/\delta})(1-u)^{\alpha-1}u^{-\beta} du. \tag{6.15}$$

3) *Assume* (6.3) *with* $1/2 < \alpha = \beta < 1$.

(i) *If* $0 \le c < \infty$ *then*

$$\lim_{t\to\infty} \mathbf{P}\{Z(t)Q(t) \le x\} = \frac{c + G_1(x)}{1+c},$$

where $G_1(x)$ *is given in* (6.14).

(ii) *If* $c = \infty$ *then* (6.15) *holds.*

4) *Let* (6.3) *holds. If* $\beta \ge 1$ *but* $\mathbf{E}\{T_1\} < \infty$, *then* (6.13) *is true.*

It is not difficult to see from (6.12) that

$$\widehat{D}_{\rho,\delta,\theta}(\infty) = 1 - \frac{r^{\theta/\delta}}{\rho^{\theta}(0)} = D_{\rho,\delta,\theta}(0) > 0.$$

This explains why it is not possible to apply Theorem 16 in this case. Hence $\beta = 1$ and $\mathbf{E}\{T_1\} = \infty$ is an open problem.

7 Continuous-State Branching Processes

In this section we discuss a continuous time and state branching process proposed by Li [10]. The author gives the following definition based on the general definition from Section 2. Suppose that G is a probability measure on $(0,\infty)$ and

$$m_I := \int_0^\infty yG(dy) < \infty.$$

Let $\{X_i : i = 0,1,2,\ldots\}$ be a family of i.i.d. exponential random variables with parameter $q > 0$ and let $\{Z_i(t) : t \ge 0, i = 1,2,\ldots\}$ be a family of i.i.d. continuous state branching processes with initial distribution G. Define another continuous state branching process $\{Z_0(t) : t \ge 0\}$ with the same transition semigroup as $\{Z_i(t) : t \ge 0, i = 1,2,\ldots\}$ but with an unspecified initial distribution. Suppose that the families $\{X_i : i = 0,1,2,\ldots\}$, $\{Z_i(t) : t \ge 0, i = 1,2,\ldots\}$ and $\{Z_0(t) : t \ge 0\}$ are independent of each other. Let $\{T_i := \inf\{t \ge 0 : Z_i(t) = 0\}$ denote the extinction time of $\{Z_i(t) : t \ge 0\}$ and $S_n := \sum_{i=0}^{n}[X_i + T_i]$. Then $S_n \to \infty$ almost surely as $n \to \infty$. Define the process $\{Z(t) : t \ge 0\}$ by

$$Z(t) = \begin{cases} Z_0(t) \text{ for } 0 \le t < T_0 + X_0; \\ Z_n(t - S_{n-1}) \text{ for } S_{n-1} \le t < S_n \text{ and } n \ge 1. \end{cases}$$

The assumptions are: (i) X_i are exponentially distributed random variables and (ii) $\{Z_i(t),\, t \ge 0\}$ are continuous time and state branching processes with the branching mechanism determined by

$$f(z) = -bz - cz^2 + \int_0^\infty (1 - e^{-zu} - zu)m(du), \quad z \ge 0,$$

where $c \ge 0$ and b are constants and $(u \wedge u^2)m(du)$ is a finite measure on $[0,\infty)$.

These assumptions imply that $\{Z(t), t \ge 0\}$ is a time homogeneous Markov process and its transition semigroup $(R_t)_{t\ge 0}$ can be computed. Li proves a continuous time and space version of the theorems of Foster [4] and Yamazato [35] using a version of Theorem 16.

Theorem 18 (Li [10]). *Assume that* $b = 0$ *and* $\sigma^2 < \infty$. *Let* $Z = (\Omega; \mathcal{G}; \mathcal{G}_t; Z(t); R_x)$ *be a realization of the semigroup* $(R_t)_{t\geq 0}$. *Then for any* $x \geq 0$ *and* $0 < y < 1$

$$\lim_{t\to\infty} R_x \left\{ \frac{\log Z(t)}{\log t} \leq y \right\} = y.$$

8 Branching Processes with Migration

The regenerative phenomena arise in a natural way in the models allowing immigration at every state of a branching process. Consider a Bienaymé-Galton-Watson branching process with immigration (BGWI). Assume that the process starts with zero ancestors $(Z(0) = 0)$ and the immigration distribution is $\mathbf{P}\{I_k = j\} = b_j, j = 0, 1, \ldots$ with $b_0 < 1$. Clearly, the process will stay at zero for time X_0 with geometric distribution, i.e., $\mathbf{P}\{X_0 = l\} = (1 - b_0)b_0^{l-1}, l = 1, 2, \ldots$ before a random number of immigrants I_0' initiate a BGWI. The process evolves until it hits the zero state at time T_0 and so on. Here we will discuss a modification of just described model, namely a branching process with migration. These processes were first studied independently by Nagaev and Khan [19], and Yanev and Mitov [38]. Later a number of authors work on these and similar models, see Yanev and Yanev [39], [40], [41], [42], Badalbaev and Yakubov [1], and Rahimov [24]. The definition below follows that given in [42].

Let $\zeta = \{\zeta_i(t), i = 1, 2, \ldots, t = 0, 1, 2, \ldots\}$ be a set of non-negative integer-valued i.i.d. r.v.'s with a p.g.f. $f(s) = \mathbf{E}\{s^{\zeta_i(t)}\} = \sum_{k=0}^{\infty} f_k s^k$. Denote $U_t(j) = \Sigma_{i=1}^{j} \zeta_i(t)$, $j = 1, 2, \ldots$, and $U_t(0) \equiv 0$. Let us also have two, independent of ζ, sets $E = \{e_1(t), e_2(t)\}$ and $I = \{I_t^+, I_t^o\}$ of i.i.d. in each set non-negative r.v. with p.g.f. $H(s_1, s_2) = \mathbf{E}\{s_1^{e_1(t)} s_2^{e_2(t)}\}$ and $G(s_1, s_2) = \mathbf{E}\{s_1^{I_t^+} s_2^{I_t^o}\}$. The branching process with (randomly controlled) migration $\{Z(t), t = 0, 1, 2, \ldots\}$ is defined by the following recurrent formula.

$$Z(0) = 0,\ Z(t) = \max\{0,\ U_t(Z(t-1)) + M_t\} \qquad (t = 1, 2, \ldots), \tag{8.1}$$

where for non-negative p, q and r $(p + q + r = 1)$ the migration component is given by $\mathbf{P}\{M_t = 0\} = q$, $\mathbf{P}\{M_t = -U_t(e_1(t) - e_2(t)\} = p$ and $\mathbf{P}\{M_t = I_t^+ \mathbb{I}_{\{Z(t-1)>0\}} + I_t^o \mathbb{I}_{\{Z(t-1)=0\}}\} = r$.

The above definition can be interpreted as follows. There are three possible scenarios for the process' evolution.

(i) Emigration with probability p: $e_1(t)$ families emigrate, i.e., $U_t(e_1(t))$ members leave the population (family emigration) and $e_2(t)$ members randomly selected from different families are also eliminated from the further evolution (individual emigration);

(ii) No migration with probability q, i.e., the reproduction is as in a BGW branching process;

(iii) State-dependent immigration with probability r: I_t^+ immigrants joint the population when it is not empty or I_t^o immigrants act as new ancestors at zero.

Notice that if $q = 1$ then $\{Z(t), t = 0, 1, 2, \ldots\}$ is a BGW process. The case $r = 1$ and $I_t^+ \equiv I_t^o$ is a BGW process with immigration. If $r = 1$ and $I_t^+ \equiv 0$ we have the Foster-Pakes model discussed in Section 2. The case $p = 1$, $e_1(t) \equiv 1$ and $e_2(t) \equiv 0$ (pure emigration) was studied by Vatutin [34].

Yanev and Yanev [39], [40] investigate the branching migration processes (8.1) under the following conditions

$$\begin{aligned} &\mathbf{E}\{\zeta_i(t)\}=1 \quad \text{(critical case)} \quad \text{and} \quad 0<Var\{\zeta_i(t)\}=2b<\infty, \\ &0<m_I^+=\mathbf{E}\{I_t^+\}<\infty, \quad 0<m_E=\mathbf{E}\{e_1(t)+e_2(t)\}<\infty, \\ &0<m_I^o=\mathbf{E}\{I_t^o\}<\infty, \\ &0\le e_1(t)\le N_1<\infty, \quad 0\le e_2(t)\le N_2<\infty \ a.s. \end{aligned} \tag{8.2}$$

The authors have shown that the following (critical) parameter

$$\Delta=\frac{rm_I^+-pm_E}{b} \tag{8.3}$$

plays a crucial role for the asymptotic behavior of the process.

Theorem 19 (Yanev and Yanev [40]). *Assume* (8.2).
(i) *If* $\Delta<0$ *then the process possesses a stationary distribution.*
(ii) *If* $\Delta=0$ *then for* $0<x<1$

$$\lim_{t\to\infty}\mathbf{P}\left\{\frac{\log Z(t)}{\log t}\le x\right\}=x.$$

(iii) *If* $\Delta>0$ *then for* $x\ge 0$

$$\lim_{t\to\infty}\mathbf{P}\left\{\frac{Z(t)}{bt}\le x\right\}=\frac{1}{\Gamma(\Delta)}\int_0^x y^{\Delta-1}e^{-y}dy,$$

where $\Gamma(\cdot)$ *is the Gamma function.*

It is not difficult to see that $Z(t)$ is an alternating regenerative process with a reflective barrier at zero. Indeed, the stay at zero X_1 is a geometric random variable with parameter $a=1-r+rg(0)$, where g(s)=G(s,1), i.e., $\mathbf{P}\{X_1=k\}=a^{k-1}(1-a),\ k=1,2,\ldots$. Leaving zero, the process jumps to a random positive level I_1^o and develops according to (8.1) until it hits zero again after some time T_1 (life period). Thus, X_1+T_1 forms the first period of regeneration. The further evolution of the process $\{Z(t),t=0,1,2,\ldots\}$ is similar with periods of regeneration $\{X_k+T_k,k=1,2,\ldots\}$, where X_k are i.i.d. as X_1 and T_k are i.i.d. as T_1.

Yanev and Yanev [41, 42] study also a BGW with (random controlled) migration stopped at zero $\{Z_1(t),t=0,1,2,\ldots\}$. This is in fact the positive part of $\{Z(t),t=0,1,2,\ldots\}$ during the first period of regeneration. Define

$$Z_1(0)=I_{X_1}^0,\ Z_1(t)=\max\{0,\ U_t(Z_1(t-1))+M_t^*\} \quad (t=1,2,\ldots), \tag{8.4}$$

where the (stopped at zero) migration is

$$M_t^*=\begin{cases} -U_t(e_1(t)-e_2(t) & \text{with probab. } p, \\ 0 & \text{with probab. } q, \\ I_t^+\mathbb{I}_{\{Z_1(t-1)>0\}} & \text{with probab. } r, \quad p+q+r=1. \end{cases} \tag{8.5}$$

Let us introduce slightly stronger moment conditions on the reproduction and immigration as follows.

$$\mathbf{E}\{I_t^+ \log(1+I_t^+)\} < \infty \quad \text{when} \quad \Delta = 0; \tag{8.6}$$

$$\mathbf{E}\{I_t^+ \log(1+I_t^+)\} < \infty, \ \mathbf{E}\{\zeta_t^2(1)\log(1+\zeta_1(t))\} < \infty \quad \text{when} \quad 0 < \Delta \leq 1. \tag{8.7}$$

For the immigration at the state zero $\{I_t^o\}$ assume a distribution which is in the normal domain of attraction of a stable law

$$g_0(s) = \mathbf{E}\{s^{I_t^o}\} = 1 - (1-s)^\theta L\left(\frac{1}{1-s}\right), \tag{8.8}$$

where $0 < \theta \leq 1$ and $L(x)$ is slowly varying at infinity. ($m_I^0 = E\{I_t^0\}$ can be finite or infinite.) In addition, assume one of the following:

$$m_I^0 = E\{I_t^0\} = \infty \text{ and } -\infty < \Delta < \infty, \tag{8.9}$$

$$m_I^0 = E\{I_t^0\} < \infty \text{ and } \Delta > 0. \tag{8.10}$$

Now we are in a position to state a conditional limit theorem for the stopped at zero processes $Z = \{T_k, Z_k(t)\}$.

Theorem 20 (Yanev and Yanev [41, 42]). *Assume* (8.2), (8.6), (8.7) *and one of* (8.9) *or* (8.10). *Then as* $\theta \to \infty$

(i) $\mathbf{P}\{T_k > t\} \sim t^{-\beta} L_1(t)$, *where*

$$\begin{array}{lll} \beta = \theta & \textit{when} & \Delta + \theta < 1, \\ \beta = \max(1-\Delta, 0) & \textit{when} & \Delta + \theta \geq 1 \ \textit{ or } (8.10) \textit{ holds}, \\ \beta = 0 \textit{ and } L_B(t) \to \text{const.} & \textit{when} & \Delta > 1. \end{array}$$

(ii) *If* $\Delta + \theta < 1$ *then*

$$\lim_{t\to\infty} \mathbf{P}\left\{\frac{Z_1(t)}{bt} \leq x | T_1 > t\right\} = H_1(x),$$

where $H_1(x)$ *has Laplace-Stieltjes transform*

$$\widehat{H}_1(\lambda) = 1 - \frac{\Gamma(1-\Delta)\Gamma(1-\theta)}{\Gamma(1-\Delta-\theta)} \frac{\lambda^\theta}{(1+\lambda)^{\Delta+\theta}} - \Delta\lambda \int_0^1 (1+x)^{-\theta}(1+x\lambda)^{-\Delta-1}dx.$$

(iii) *If* $\Delta + \theta \geq 1$ *or* (8.10) *holds, then*

$$\lim_{t\to\infty} \mathbf{P}\left\{\frac{Z_1(t)}{bt} \leq x | T_1 > t\right\} = H_2(x),$$

where $H_2(x)$ *has Laplace-Stieltjes transform* $\widehat{H}_2(\lambda) = (1+\lambda)^{-\beta}$.

In the end of this section we will present results for a generalization of the process with migration (8.1). Namely, consider the alternating regenerative process from Section 6 with life-cycles given by the stopped-at-zero processes with migration $Z = \{T_k, Z_k(t)\}$ and zero-cycles $\{X_k\}$ with c.d.f. $A(x)$ which is not necessary geometric as in (8.1). The asymptotic results for $\mathbf{P}\{T_1 > t\}$ and the conditional limiting distributions for $\{Z_1(t), t \geq 0\}$ obtained in Theorem 41 above guarantee that **Condition 3** and **Condition 4** of Theorem 16 hold. Therefore, we can apply Theorem 16 to this modification of $Z(t)$ with non-geometric zero-cycles. Next three theorems present some recent results in the critical case. Recall the assumption (6.1) from Section 6 that X_k and T_k have non-lattice distributions with $P\{X_k = 0\} = P\{T_k = 0\} = 0$ and

$$\lim_{t\to\infty} \frac{P\{X_k > 0\}}{P\{T_k > 0\}} = c \qquad (0 \leq c \leq \infty). \tag{8.11}$$

Theorem 21 (Yanev et al. [37]). *Assume* $\theta = 1$, $\Delta \leq 0$ *and* $\mathbf{E}\{I_t^+ \log(1 + I_t^+)\} < \infty$.
(i) *If* $\mathbf{E}X_k < \infty$ *then*

$$\lim_{t\to\infty} \mathbf{P}\left\{\frac{\log Z(t)}{\log t} \leq x\right\} = x \qquad (0 \leq x \leq 1).$$

(ii) *If* $\mathbf{E}X_k = \infty$ *and* $\mathbf{P}\{X_k > 0\} = t^{-\alpha}L(t)$ *with* $1/2 < \alpha < 1$, *then*

$$\lim_{t\to\infty} \mathbf{P}\left\{\frac{\log Z(t)}{\log t} \leq x | Z(t) > 0\right\} = x \qquad (0 \leq x \leq 1). \tag{8.12}$$

(iii) *If* $\mathbf{E}X_k = \infty$, $\mathbf{P}\{X_k > 0\} = t^{-1}L(t)$ *and in* (8.11) $c = \infty$ *then* (8.12) *is true.*
(iv) *If* $\mathbf{E}X_k = \infty$, $\mathbf{P}\{X_k > 0\} = t^{-1}L(t)$ *and in* (8.11) $0 \leq c < \infty$ *then*

$$\lim_{t\to\infty} \mathbf{P}\left\{\frac{\log Z(t)}{\log t} \leq x\right\} = \frac{c}{1+c} + \frac{x}{1+c}, \quad 0 \leq x \leq 1.$$

Theorem 22 (Yanev et al. [37]). *Assume* $\theta + \Delta \geq 1$, $0 < \Delta < 1/2$ *and* (8.7).
(i) *If* $0 \leq c < \infty$ *then for* $x \geq 0$

$$\lim_{t\to\infty} \mathbf{P}\left\{\frac{Z(t)}{bt} \leq x\right\} = \mathbf{P}\{Z_\infty^{(1)} \leq x\},$$

where $\mathbf{E}\{Z_\infty^{(1)}\} = \theta$ *and*

$$\mathbf{P}\left\{Z_\infty^{(1)} \leq x\right\} = 1 - \frac{1}{\mathrm{B}(\Delta, 1-\Delta)} \int_0^1 e^{-x/u} u^{\Delta-1} (1-u)^{-\Delta} du$$

(ii) *If* $c = \infty$ *then for* $x \geq 0$

$$\lim_{t\to\infty} \mathbf{P}\left\{\frac{Z(t)}{bt} \leq x | Z(t) > 0\right\} = \mathbf{P}\{Z_\infty^{(2)} \leq x\},$$

where $\mathbf{E}\{Z_\infty^{(2)}\} = \Delta/(\Delta + \alpha)$ *and*

$$\mathbf{P}\{Z_\infty^{(2)} \leq x\} = 1 - \frac{1}{\mathrm{B}(\Delta, \alpha)} \int_0^1 e^{-x/u} u^{\Delta-1} (1-u)^{\alpha-1} du$$

Theorem 23 (Yanev et al. [37]). *Assume* $\theta+\Delta<1$, $1/2<\theta<1$, (8.6) *and* (8.7).

(a) *Assume* $\mathbf{E}X_k<\infty$ *or* $\mathbf{E}X_k=\infty$ *and* $P\{X_k>t\}=t^{-\alpha}L(t)\quad(1/2<\alpha\leq 1)$. *If* $\theta<\alpha$ *then for* $x\geq 0$

$$\lim_{t\to\infty}\mathbf{P}\left\{\frac{Z(t)}{bt}\leq x\right\}=G_1(x). \tag{8.13}$$

The limiting c.d.f. $G_1(x)$ *has Laplace-Stieltjes transform*

$$1-\frac{1}{\mathrm{B}(1-\theta,\theta)}\left[D\lambda^{\theta}\int_0^1 g(u)(1-u)^{\theta-1}du-\Delta\lambda\int_0^1\int_0^1 h(u,y)(1-u)^{\theta-1}dydu\right]$$

where $D=\Gamma(1-\Delta)\Gamma(1-\theta)/\Gamma(1-\Delta-\theta)$, $g(u)=(1+\lambda u)^{-\Delta-\theta}$ *and* $h(u,y)=(1+\lambda yu)^{-\Delta-1}u^{1-\theta}$.

(b) *Assume* $\mathbf{E}X_k=\infty$ *and* $P\{X_k>t\}=t^{-\alpha}L(t)\quad(1/2<\alpha\leq 1)$. *If* $\theta>\alpha$ *then for* $x\geq 0$

$$\lim_{t\to\infty}\mathbf{P}\left\{\frac{Z(t)}{bt}\leq x|Z(t)>0\right\}=G_2(x). \tag{8.14}$$

The limiting c.d.f. $G_2(x)$ *has Laplace-Stieltjes transform*

$$1-\frac{1}{\mathrm{B}(1-\theta,\alpha)}\cdot\left[D\lambda^{\theta}\int_0^1 g(u)(1-u)^{\alpha-1}du-\Delta\lambda\int_0^1\int_0^1 h(u,y)\frac{(1-u)^{\alpha-1}}{(1-y)^{\theta}}dydu\right]$$

where D, $g(u)$ *and* $h(u,y)$ *are as in part(a).*

(c) *Assume* $\mathbf{E}X_k=\infty$ *and* $P\{X_k>t\}=t^{-\theta}L(t)$. (i) *If in* (8.11) $0\leq c<\infty$ *then*

$$\lim_{t\to\infty}\mathbf{P}\left\{\frac{Z(t)}{bt}\leq x\right\}=\frac{c}{1+c}+\frac{G_1(x)}{1+c},\quad x\geq 0,$$

where $G_1(x)$ *is from* (8.13). (ii) *If in* (8.11) $c=\infty$ *then* (8.14) *is true.*

Corollary 1 (Yanev et al. [37]). *The limiting distributions* (8.13) *and* (8.14) *belong to the normal domain of attraction of a stable law with parameter* $1/2<\theta<1$.

9 Time Non-Homogeneous State-Dependent Immigration

In this section we discuss briefly another extension of the Foster-Pakes and Yamazato models. The new assumption is that the distribution of the immigration at zero varies with the time. This leads to a class of branching processes which are non-homogeneous Markov chains and cannot be inscribed in the general definition in Section 2. These processes are studied in [12] (discrete time) and [13] (continuous time). They exhibit different limiting behavior depending on the rate of change of the immigration at zero. Below we present some interesting limit results for continuous time processes in the critical case. Define the time non-homogeneous Markov chain $\{Z(t),t\geq 0\}$ with transition probabilities as $h\to 0$

$$\mathbf{P}\{Z(t+h)=j|Z(t)=i\}=\begin{cases}\delta_{0i}+q_i(t)h+o(h) & i=0,j\geq 0,\\ \delta_{ij}+ip_j+1-ih+o(h) & 1\leq i\leq j+1,\\ o(h) & 0\leq j<i-1.\end{cases}$$

where $q_0(t) < 0, q_j(t) \geq 0\ (j \geq 1), \sum_{j=0}^{\infty} q_j(t) = 0\ (t \geq 0), p_1 < 0, p_j \geq 0\ (j \neq 1), \sum_{j=0}^{\infty} p_j = 0$. Set for simplicity $Z(0) = 0\ a.s.$

Assume that

$$\sum_{j=1}^{\infty} jp_j = 0, 0 < 2b = \sum_{j=1}^{\infty} j(j-1)p_j < \infty \quad (critical\ \ case) \tag{9.1}$$

$$0 < m_I(t) = \sum_{j=1}^{\infty} jq_j(t) \to 0, \quad c_I(t) = \sum_{j=2}^{\infty} j(j-1)q_j(t) \to 0 \tag{9.2}$$

and

$$\sup_{t \geq 0} m_I(t) < \infty \quad \sup_{t \geq 0} c_I(t) < \infty. \tag{9.3}$$

As the following theorem reveals, $\{Z(t)\}$ has different limiting behavior depending on the rate of convergence in (9.2).

Theorem 24 (Mitov et al.[13]). *Assume* (9.1), (9.2), (9.3),

$$m_I(t) = L(t)t^{-\theta} \quad and \quad c_I(t) = o(m_I(t)\log t) \quad (t \to \infty),$$

where $0 \leq \theta < 1$, *and* $L(.)$ *is slowly varying at infinity.*

(a) *If* $0 < \theta < 1$ *or* $\{\theta = 0$ *and* $0 \leq \lim_{t \to \infty} b^{-1}L(t)\log t < 1\}$ *then*

$$\lim_{t \to \infty} \mathbf{P}\left\{\frac{\log Z(t)}{\log t} \leq x | Z(t) > 0\right\} = x, \qquad (0 < x < 1).$$

(b) *If* $\theta = 1$ *and* $\lim_{t \to \infty}(L(t)\log t)\left(\int_0^t m_I(u)du\right)^{-1} = a, \ (0 \leq a \leq \infty)$, *then*

(i) *for* $0 < x < 1$

$$\lim_{t \to \infty} \mathbf{P}\left\{\frac{\log Z(t)}{\log t} \leq x | Z(t) > 0\right\} = \frac{ax}{1+a};$$

(ii) *for* $x \geq 0$

$$\lim_{t \to \infty} \mathbf{P}\left\{\frac{Z(t)}{bt} \leq x | Z(t) > 0\right\} = \frac{a}{1+a} + \frac{1-e^{-x}}{1+a}.$$

(c) *If* $\int_0^{\infty} m_I(t)dt < \infty$ *and* $c(t) = o(t^{-1})$, *then for* $x > 0$

$$\lim_{t \to \infty} \mathbf{P}\left\{\frac{Z(t)}{bt} \leq x | Z(t) > 0\right\} = 1 - e^{-x}.$$

The above results show that if the intensity of the immigration at zero decreases to zero then only conditional limit distributions are possible. Moreover, the non-degenerate sample paths are of two different types: (i) $Z(t) \sim t^{\xi}$ with probability $a/(1+a)$ where ξ is uniformly distributed in $[0,1]$ and (ii) $Z(t) \sim \eta t$ with probability $1/(1+a)$ where η has an exponential distribution.

10 Concluding Remarks

The classical branching models without immigration (as Bienaymé-Galton-Watson processes, Markov branching processes, Bellman-Harris processes and others with absorbing zero state) successfully describe the dynamics of an population without interaction with the surrounding environment. The regenerative branching processes (as particular case of the alternating regenerative processes) can be viewed as models of "branching stochastic periodicity" modeling specific interactions with the environment. This make them more flexible an adequate as a modeling tool in studying real-world phenomena is such diverse areas as economy, climatology, epidemiology, among others.

The immigration at zero mechanism allows one to construct a regenerative process using a classical branching models as its life-cycles. Our fundamental Theorem16 can be used to study the limiting behavior of such a regenerative process. We illustrated this in Sections 6-8.

References

[1] Badalbaev, I.S.; Yakubov, T. D. Limit theorems for critical Galton-Watson branching processes with migration. *Theory Probab. Appl.* 1995, *40*, 599–612.

[2] Dynkin, E. B. Limit theorems for sums of independent random variables. *Izv. Acad. Nauk SSSR* 1955, *19*, 247–266.

[3] Erickson, K. B. Strong renewal theorems with infinite mean. *Trans. Amer. Math. Soc.* 1970, *151*, 263–291.

[4] Foster, J. A limit theorem for a branching process with state dependent immigration. *Ann. Math. Statist.* 1971, *42*, 1773–1776.

[5] Goldstein, M. I. Critical age-dependent branching processes: Single and multitype. *Z. Wahrscheinlichkeitstheorie und Verw. Gebiete* 1971, *17*, 74–88.

[6] Feller, W. *An Introduction to Probability Theory and its Applications* . Wiley: New York, 1971, Vol. 2.

[7] Kalashnikov V. *Topics on Regenerative Processes*. CRC Press: Boca Raton, 1994.

[8] Kulkarni, M. V.; Pakes, A. G. The total progeny of a simple branching process with state-dependent immigration. *J. Appl. Prob.* 1983, *20*, 472–481.

[9] Lamperti, J. A contribution to renewal theory. *Proc. Amer. Math. Soc.* 1961, *12*, 72–731.

[10] Li, Z. H. Asymptotic behavior of continuous time and state branching processes. *J. Austral. Math. Soc. Ser. A)* 2000, *68*, 68–84.

[11] Mitov, K. V. Limit theorems for regenerative excursion processes. *Serdica - Bulg. Math. J.* 1999, *25*, 19–40.

[12] Mitov, K. V. and Yanev, N. M. Critical Galton-Watson branching processes with decreasing state-dependent immigration. *J. Appl. Prob.* 1984, *21*, 22–39.

[13] Mitov, K. V.; Vatutin, V. A.; Yanev, N. M. Continuous-time branching processes with decreasing state-dependent immigration. *Adv. Appl. Prob.* 1984, *16*, 697–714.

[14] Mitov, K. V.; Yanev, N. M. Bellman-Harris branching processes with state-dependent immigration. *J. Appl. Prob.* 1985, *22*, 757–765.

[15] Mitov, K. V.; Yanev, N. M. Bellman-Harris branching processes with a special type of state-dependent immigration. *Adv. Appl. Prob.* 1989, *21*, 270–283.

[16] Mitov, K. V.; Yanev, N. M. Limit theorems for alternating renewal processes in the infinite mean case. *Adv. Appl. Prob.* 2001, *33*, 896–911.

[17] Mitov, K. V.; Yanev, N. M. Regenerative processes in the infinite mean cycle case. *J. Appl. Prob.* 2001, *38*, 165–179.

[18] Mitov, K. V.; Yanev, N. M. Critical Bellman-Harris branching processes with infinite variance allowing state-dependent immigration. *Stochastic Models* 2002, *18*, 2:281–300.

[19] Nagaev, S. V.; Khan, L. V. Limit theorems for Galton- Watson branching processes with migration. *Theory Probab. Appl.* 1980, *25*, 523–534.

[20] Nakagawa, T.; Sato, M. A Galton-Watson process with state-dependent immigration. *Res. Rep. Nagaoka Tech. Coll.* 1974, *9*, 4:177–182.

[21] Pakes, A. G. A branching process with a state-dependent immigration component. *Adv. Appl. Prob.* 1971, *3*, 301–314.

[22] Pakes, A. G. Some results for non-supercritical Galton-Watson processes with immigration. *Math. Biosci.* 1975, *24*, 71–92.

[23] Pakes, A. G. On the age distribution of a Markov chain. *Adv. Appl. Prob.* 1978, *15*, 65–77.

[24] Rahimov, I. *Random sums and branching stochastic processes* . Lecture Notes in Statistics, Springer, New York, 1995.

[25] Sigman, K.; Wolff, R. A review of regenerative processes. *SIAM Review* 1993, *35*, 269–288.

[26] Slack, R. S. A branching process with mean one and possibly infinite variance. *Z. Wahrscheinlichkeitstheorie und Verw. Gebiete* 1968, *9*, 139–145.

[27] Slavtchova, M. N.; Yanev, N. M. Convergence in distribution of supercritical Bellman-Harris branching processes with state-dependent immigration. *Mathematica Balkanica (New Series)* 1990, *4*, 1:35–42.

[28] Slavtchova, M. N.; Yanev, N. M. Non-critical Bellman- Harris branching processes with state-dependent immigration. *Serdica Bulg. Math. J.* 1991, *17*, 67–79.

[29] Vatutin, V. A. Limit theorems for critical Bellman – Harris branching processes with infinite variance. *Theory Probab. Appl*. 1976, *21*, 861–863 (in Russian).

[30] Vatutin, V. A. Discrete limit distributions for the number of particles in critical Bellman–Harris branching processes. *Theory Probab. Appl.* 1977, *22*, 150–155 (in Russian).

[31] Vatutin, V. A. New limit theorem for critical Bellman-Harris branching process. *Math. Sbornik* 1979, *109*, 440–452 (in Russian).

[32] Vatutin, V. A. On a class of limit theorems for critical Bellman-Harris branching process. *Theory Probab. Appl*. 1981, *26*, 818–823 (in Russian).

[33] Vatutin, V. A. Branching processes with infinite variance. In *Proc. 4th International Summer School on Probability Theory and Math. Statistics, Obretenov* , A. Editor, Bulg. Acad. Sci.: Sofia, Bulgaria, 1983, 9–38 (in Russian).

[34] Vatutin, V. A. A critical Galton-Watson branching process with emigration. *Theory Probab. Appl*. 1977, *22*, 482–497.

[35] Yamazato, M. Some results on continuous time branching processes with state-dependent immigration. *J. Math. Soc. Japan.* 1975, *27*, 479–496.

[36] Yanev, G. P.; Mitov, K. V.; Yanev, N. M. Critical branching regenerative processes with migration. *J. Applied Statistical Science*, 2003, *12*, 1:41–54.

[37] Yanev, N. M.; Mitov, K. V.; Yanev, G. P. Renewal, regenerative, and branching processes with stable distributions. *J. Mathematical Sciences*, 2004, *122*, 4:3438–3448.

[38] Yanev, N. M.; Mitov, K. V. Controlled branching processes: the case of random migration. *C.R. Acad. Bulg. Sci*. 1980, *33*, 4:433–435.

[39] Yanev, G. P.; Yanev, N. M. Critical branching processes with random migration. In Heyde, C. C. (Eds.) *Branching Processes: Proceedings of the First World Congress, Lecture Notes in Statistics*, Springer: New York, 1995, Vol. 99, 36–45.

[40] Yanev, G. P.; Yanev, N. M. Branching processes with two types of emigration and state dependent immigration. In Heyde, C. C.; Prohorov, Yu. V.; Pyke, R.; Rachev, S. T. (Eds.) *Athens Conference on Applied Probability and Time Series Analysis, Volume 1: Applied Probability, Lecture Notes in Statistics* , Springer, New York, 1996, Vol. 114, 216–228.

[41] Yanev, G. P.; Yanev, N. M. Limit theorems for branching processes with random migration stopped at zero. In Athreya K. B.; Jagers P. (Eds.) *Classical and Modern Branching processes, IMA volumes in Mathematics and Its Applications* , Springer: York, 1997, 323–336.

[42] Yanev, G. P.; Yanev, N. M. Limit theorems for branching processes with random migration components. *Pliska Stud. Math. Bulgar*. 2000, *13*, 199–205.

[43] Zolotarev, V. M. More exact statements of several theorems in the theory of branching processes. *Theory Probab. Appl*. 1957, *2*, 256–265.

In: Records and Branching Processes
Editors: M. Ahsanullah, G. P. Yanev, pp. 63-103
ISBN 978-1-60456-526-3

Chapter 4

CONDITIONAL LIMIT THEOREMS FOR CONTINUOUS TIME AND STATE BRANCHING PROCESS

Anthony G. Pakes*
School of Mathematics and Statistics
University of Western Australia
Crawley, WA 6009
Australia

Abstract

This paper gives a unified treatment of conditional limit theorems for the non-supercritical continuous time and state branching process, the CB-process $(Z_t; t \geq 0)$. Most previous work looks separately at the cases where the zero state either is, or is not, accessible. Here limit theorems for Z_t are found by conditioning on $Z_{t+\tau} > \varepsilon$ where τ may depend on t. When the zero state is accessible, in time T say, results are given for conditioning on events such as $\{t+\tau < T < \infty\}$.

A stationary measure is shown to exist iff extinction is possible. In the critical case this measure determines a limit law for Z_t given that $T \in t + A$, where A has finite Lebesgue measure.

When $P(T = \infty) = 1$ a surrogate notion of extinction is the last time $\mathcal{L}_\varepsilon$ at which (Z_t) jumps by more than $\varepsilon > 0$. The law of $\mathcal{L}_\varepsilon$ is found as well as some limit theorems for Z_t conditioned on $\{t+\tau < \mathcal{L}_\varepsilon < \infty\}$.

Key words and phrases. Branching process, continuous time and state, Lévy process, limit theorems, extinction, stationary measure.

1 Introduction

A discrete-state Markovian branching process (BP) is a Markov chain (X_t) indexed by $t = 0, 1, \ldots$ (the simple BP) or by $t \geq 0$ (the Markov BP or MBP) and satisfying the branching property

$$P(X_t = j \,|X_0 = i) = P^{*i}(X_t = j \,|X_0 = 1) \quad (i, j = 0, 1, \ldots).$$

*E-mail address: pakes@maths.uwa.edu.au

Thus a BP models growth of a population in which individuals reproduce independently of each other according to a non-degenerate offspring law. Let m denote the per-capita mean number of offspring. If $m \leq 1$ then the absorbing zero state is attained, that is, extinction occurs at a finite time T. Central results of branching process theory are limit theorems for the population size, conditioned on extinction having not yet occurred. Their original forms for the simple BP are due to Yaglom (1947), with generalizations and variations since then. See Asmussen and Hering (1983) for a monograph account, and Pakes (1999) for a recent discussion and extensions. Corresponding results for the MBP quickly followed (Sevast'yanov (1951)), and later Conner (1967) showed that embedding arguments can be used to transfer limit theorems for the simple BP to the MBP case.

In this paper we are concerned with the continuous time and state counterpart of the above classical models, with a principal aim of reviewing and unifying its limit theory. Previous studies are largely concerned with broad subclasses, and hence it is useful to have on record a complete treatment. We begin with a short survey, and expand it a little in §2 after introducing basic concepts and notation.

Branching processes taking values in $[0,\infty)$ were introduced by Jĭrina (1958), for both discrete and continuous time. Constructions of the discrete-time version, called the Ĭrina process (JP), in terms of subordinators were published independently by Pakes (1979) and Kallenberg (1979). The continuous-time version, which we call the CB-process, was earlier constructed by Lamperti (1967) as a randomly time transformed spectrally positive Lévy process. We recall this construction in §7 where we first use it. A physical picture for the branching property of these processes is a continuous randomly fluctuating mass which has the property that disjoint subsets evolve in time independently of each other. In this context, we let m denote the expected mass produced after one unit of time by one unit of initial mass.

A key difference between discrete-state and continuous-state BP's is that in the latter case it is possible for the zero state to be unattainable when $m \leq 1$. In this case the population mass is always positive, though it tends almost surely to zero. For example, denote the population mass size process for the JP by $\{\eta_n\}$. Seneta and Vere-Jones (1969) proved that there is a sequence of constants $0 < c_n \to \infty$ such that $\{c_n\eta_n\}$ has a non-degenerate limiting law iff the offspring distribution function $P(\eta_1 \leq x|\eta_0 = 1)$ has a left-extremity $\ell > 0$. They assumed that $m < 1$, but this is is unnecessary for their proof. Subsequently it was observed that the convergence holds almost surely; see Pakes (1979) for sources. Kallenberg (1979) subsequently obtained the same result – he doesn't cite Seneta and Vere-Jones (1969). Grey (1974) proved a corresponding result for the CB-process.

When $m < 1$ and without any restriction on ℓ, Seneta and Vere-Jones (1968) proved that for each $\varepsilon > 0$,

$$\lim_{n\to\infty} P(\eta_n \leq x|\eta_n > \varepsilon) \tag{1.1}$$

exists as an honest and non-degenerate law on $[\varepsilon,\infty)$. This may be considered to be an analogue of the Yaglom theorem for the subcritical simple BP. They observed that (1.1) is true for $\varepsilon = 0$ if the offspring law assigns positive probability to the origin, a case first considered by Buhler (1968) under stronger moment conditions.

The critical case ($m = 1$) counterpart of the Yaglom theorem for the JP is due in the first instance to Shurenkov (1973), and completed by Seneta (1976). They show when

$$\sigma^2 = Var(\eta_1|\eta_0 = 1) < \infty$$

that for each $\varepsilon > 0$,

$$\lim_{n\to\infty} P\left(\eta_n \leq (\sigma^2/2)nx|\eta_n \geq \varepsilon\right) = 1 - e^{-x}. \tag{1.2}$$

The corresponding situation for the CB process is more fragmented, and hence we describe it in the next section.

The conditioning events in (1.1) and (1.2), and the continuous-time version we will use below are, in one respect, mere technical devices for inducing general limit theorems. However, they, and others like it, can be construed in a population modelling context as conditioning on the event that the current population size is above a critical threshold below which the population is deemed to be at risk. This derives from ecological field studies indicating that populations which fall to very small sizes become susceptible to environmental disturbances and genetic deterioration that greatly increase their likelihood of extinction. See Ginzberg et al. (1982) for the first ecological formulation of this idea, and Morris and Doak (2002) for extended discussion.

The main results of the paper are summarized at the end of §2, and we end this introduction with a brief account of other occurrences of CB-processes, denoted here by$(Z_t : t \geq 0)$. They stand in the same relationship to simple BP's, as do infinitely divisible laws to the limit theory for sums of independent random variables. First, it is clear that the fundamental branching property and the continuity of the state space imply that, for each$t$, Z_t has an infinitely divisible law. Next, CB-processes arise as the limit processes of rescaled simple BP's. For $k = 1, 2, \ldots$, let $(X_n^{(k)} : n = 0, 1, \ldots)$ denote a simple BP with $X_0^{(k)} = x_k$ and whose offspring law may depend on k. If there is a sequence of constants $b_k \uparrow \infty$ such that the finite-dimensional laws of the process $(X_{[kt]}^{(k)}/b_k : t \geq 0)$ converge to those of a limiting process, then that limit is a CB-process. Conversely any CB-process can be obtained in this way. As a general proposition this is due to Lamperti (1967), but see Grimvall (1974) for optimal results in this direction. Particular cases were known earlier. Le Gall (1999, §II.1) gives the most detailed monograph account, at least for the non-supercritical case. He shows how the branching property induces the characteristic exponent, or cumulant function (denoted cf), of a spectrally positive Lévy process (see (2.1)). Conversely, starting with such a cf allows the construction of a family of MBP's whose transition laws converge weakly to those of a CB-process.

CB-processes arise in certain constructions of stochastic analysis, Bertoin (1997) and Warren (1997) for example. Recent work is concerned with defining genealogical structure of CB-processes in terms of a local time for the generating Lévy process, and also as a weak limit of the more tangible genealogies associated with approximating simple BP's. See Duquesne and Le Gall (2002), and Lambert (2003), for a recent account of this work. Harn and Steutel (1993), and Pakes (1997b), discuss certain questions about properties of infinitely divisible laws and Poisson mixtures which can be answered in terms of CB-processes. The CB-process constructed from a compound Poisson process has been used to model the fluctuating levels of rabies virus in an infected host. See Pakes and Trajstman (1985), Kim (1998), and their references.

Finally, we mention that Lambert (2008) discusses quasi-stationary and limiting conditional laws for the subcritical CB-process with a more probabilistic twist than ours. He restricts attention to the case where the zero state is accessible.

2 The CB-Process

We need the following notation and concepts. As we have mentioned, the Lamperti construction of a CB-process $(Z_t : t \geq 0)$ is a randomly time changed spectrally positive Lévy process (L_t) whose cumulant function (cf) is

$$\begin{aligned}\psi(s) &= -\log E(e^{-s(L_1-L_0)}) \\ &= as - \nu s^2/2 + \int_{(0,1]} (1-e^{-sx}-sx)Q(dx) + \int_{(1,\infty)} (1-e^{-sx})Q(dx).\end{aligned} \tag{2.1}$$

Here $a \in I\!R$ is the deterministic linear drift rate, $\nu \geq 0$ is the variance rate of the Brownian component, and Q is a Lévy measure, that is, a measure on $[0,\infty)$ satisfying $\int_0^\infty (1 \wedge x^2)Q(dx) < \infty$. We call (a,ν,Q) the defining triple of the CB-processes. The cf $\Psi(s,t)$ of Z_t is obtained from the integrated backward equation

$$\int_s^{\Psi(s,t)} du/\psi(u) = t. \tag{2.2}$$

In other words,

$$E_z\left(e^{-sZ_t}\right) = e^{-z\Psi(s,t)}, \tag{2.3}$$

where we use the notation $E_z(\cdot) = E(\cdot|Z_0 = z)$. Similarly, define $P_z(\cdot) = P(\cdot|Z_0 = z)$, and observe that applying the monotone convergence theorem to (2.3) shows that the probability of extinction by time t is

$$P_z(Z_t = 0) = e^{-z\Psi(t)}, \tag{2.4}$$

where $\Psi(t) = \lim_{s\to\infty}\Psi(s,t)$. If $Q \equiv 0$ then the CB-process is the diffusion process satisfying the Ito equation $dZ_t = -aZ_t dt + \sqrt{\nu Z_t} dB_t$ where (B_t) is a standard Brownian motion process.

A CB-process is classified as subcritical, critical, or supercritical according as its drift parameter

$$D = \psi'(0+) = a + \int_{1+}^{\infty} xQ(dx) \tag{2.5}$$

is negative, zero or positive, respectively, and we observe that $-\infty < D \leq \infty$. It is easily shown that the mean population size by time t grown from an initial size of unity is

$$E_1(Z_t) = e^{Dt},$$

and hence we interpret D as the mean instantaneous growth rate. In addition $m = e^D$. If D is finite then (2.1) can be rewritten as

$$\psi(s) = Ds - \nu s^2/2 + \int_0^\infty (1 - e^{-sx} - sx)Q(dx). \tag{2.6}$$

There are three important subcases.

Case 1. $q_1 := \int_0^1 xQ(dx) < \infty$, i.e. the jump component of (L_t) is a subordinator.

Case 2. $q_2 := Q((0,1]) < \infty$, i.e. the jump component of (L_t) is a compound Poisson process.

Case 3. $q_1 = \infty$.

In the subordinator case we can write

$$\psi(s) = -\kappa s - \nu s^2/2 + \int_0^\infty (1 - e^{-sx})Q(dx), \tag{2.7}$$

where $\kappa = q_1 - a$, and then

$$D = -\kappa + \int_0^\infty xQ(dx).$$

If $n(\cdot)$ is the Lévy measure of a subordinator, then $\widetilde{n}(s) = \int_0^\infty (1 - e^{-sx})n(dx)$ denotes the corresponding cf, and if $n(\cdot)$ is a finite measure, then $\widehat{n}(s) = \int_0^\infty e^{-sx}n(dx)$ denotes its Laplace–Stieltjes transform.

We confine our discussion to the conservative case $P_x(Z_t < \infty) \equiv 1$, equivalent to the condition

$$\int_0^\varepsilon ds/|\psi(s)| = \infty \quad (\varepsilon > 0).$$

This can be violated only if $D = \infty$, an extreme supercritical configuration. Limit theory for the supercritical case, amongst other topics, is comprehensively treated by Bingham (1976). Zhang (1983) is an independent treatment of the basic limit theorem for this case.

If $D \leq 0$, then similarly to the JP, the zero state may not be accessible to CB-process sample paths. Grey (1974) proved that the zero state *is* accessible iff

$$\int_1^\infty ds/\psi(s) > -\infty. \tag{2.8}$$

If this condition holds then the extinction time $T = \inf\{t > 0 : Z_t = 0\}$ is almost surely finite. For Case 1, (2.8) holds iff $\nu > 0$ (Pakes & Trajstman (1985)). These authors were motivated by Case 2 with no Brownian component ($\nu = 0$) where sample paths eventually decay

exponentially fast to zero. This motivated them to seek limit theorems for Z_t conditional on $Z_t > \varepsilon > 0$. No additional moment assumptions were required when $D < 0$, and only a mild regular variation condition was assumed when $D = 0$. Their critical case result is a counterpart for Zolotarev's (1957) Theorem 7 for the MBP, or Slack's (1968) theorem for the simple BP.

Condition (2.8) holds if $\nu > 0$, and it may also hold for Case 3 and $\nu = 0$. We show this using Example 3.1 in the next section. For $D \leq 0$ the accessible-zero case has recently been considered by Li (2000). He obtains direct analogues of the Yaglom theorems by conditioning on non-extinction. He makes no further moment assumption for the case $D < 0$, and for $D = 0$ he assumes $\int_0^\infty x^2 Q(dx) < \infty$, effectively an assumption of finite offspring variance.

A principal aim of this paper is to provide general results which hold whether or not extinction can occur. Our proofs are simple and make no unnecessary moment assumptions. Behaviour near the origin of the distribution of Z_t is the focus of §3. Example 3.1 provides a scale of comparison which indicates the point at which (2.8) fails. In the case it does fail, Theorem 3.1 describes behaviour at the left extremity of the distribution function (DF) of Z_t.

Section 4 deals with the critical case $D = 0$ under the mild growth condition (4.2) used by Bingham (1976). Theorem 4.1 is a limit theorem for Z_t, given that $Z_t > \varepsilon$ where ε is a positive constant. This result holds independently of (2.8). Theorem 4.2 shows that if (2.8) does hold, then we can allow $\varepsilon = 0$ in Theorem 4.1, i.e., we can condition on $T > 0$. Theorem 4.3 is a representation result for the limit law. If $\int_1^\infty x^2 Q(dx) < \infty$, then the limit law is negative-exponential. Theorems 4.4 – 4.10 give various extensions of the basic Theorems 4.1 and 4.2. In particular, Theorems 4.4 and 4.9, respectively, generalize Theorems 5.2 and 5.3 of Li (1999), and Theorem 4.9 interpolates the 'classical type' Theorems 4.1 and 4.8.

In §5 we show when $D \leq \infty$ that a stationary measure for the CB-process exists iff the origin is accessible ((2.8) holds), and in case that it is, then the stationary measure is unique. Theorem 5.1 retraces a prior (and independent) result of Ogura (1970, Lemma 1.2), but with differing detail. In the case $D \leq 0$ our proof yields a representation, (5.7) below, which shows that the stationary measure is absolutely continuous. For the critical case $D = 0$, Theorem 5.2 is a ratio-limit theorem for the (totally finite) spectral measure of Z_t, and Theorem 5.3 is a conditional limit theorem, given extinction "in the near future", which is valid with no further moment assumptions.

The subcritical case $D < 0$ is discussed further in §6. Theorems 6.1 and 6.2 are ratio limit results expressed in terms of a $(-D)$–invariant measure for the transition semigroup. Theorem 6.2 also asserts a fundamental conditional limit theorem. These results are then used to prove other limit theorems for Z_t conditioned by events implying that $Z_t > 0$. In the case that (2.8) holds, Theorem 6.6 is the analogue of the Yaglom theorem for the subcritical simple BP. This result is essentially the same as Theorem 4.3 of Li (1999), but its proof differs in that it is effectively a corollary of the more general Theorem 6.2. Theorem 6.7 is an extension of this basic result. Theorem 6.8 identifies a unique one-parameter family of quasi-stationary laws, and conditions for their existence. This result, in essence, is the analogue of known results for the simple BP and the MBP.

We have suggested above that the event $\{Z_t > \varepsilon\}$ captures the idea of a population not in immediate extreme risk of extinction. From a modelling point of view, however, the complementary event $\{Z_t \leq \varepsilon\}$ is not entirely satisfactory as a conceptualization of "extinction", because it does not preclude the possibility of later large population sizes. One may ask if there are alternative concepts of extinction. The answer is 'yes' when $\nu = 0$ and $q_2 < \infty$. By explicitly carrying out Lamperti's (1967) construction, Pakes and Trajstman (1985) showed there is a last positive jump, at time T_L say, after which the population size decays exponentially fast toward zero with no further stochastic activity. Thus it is reasonable to define T_L as the time of extinction in this case. Pakes (1988) obtained a limit heorem for $Z(T_L)$, given that $T_L > t$. This result can be extended as follows in ways similar to those in §§4,6.

If $q_2 = \infty$ there is no last jump – stochastic activity never ceases. However, in the spirit of conditioning on $Z_t > \varepsilon$, if $0 < q_2 \leq \infty$ then there is a random time $\mathcal{L}_\varepsilon$ at which there is a jump exceeding ε and after which there is no such jump. Clearly $\mathcal{L}_\varepsilon \leq T_L$ if $q_2 < \infty$. Thus local stochastic fluctuations following the time $\mathcal{L}_\varepsilon$ are relatively small, although their aggregate effect could allow the population size to exceed ε subsequent to $\mathcal{L}_\varepsilon$. We will accept that $\mathcal{L}_\varepsilon$ provides a notion of extinction when $q_2 > 0$. In §7 we compute the law of $\mathcal{L}_\varepsilon$ and give some representative conditional limit theorems in terms of events such as $\{t + \tau < \mathcal{L}_\varepsilon < \infty\}$. These results are valid even if (2.8) holds. Theorem 7.4 generalizes the above cited result in Pakes (1988).

The results we prove for $D < 0$ embrace similar results for the supercritical case $D > 0$. To see this, recall the dichotomy

$$P_z\{\lim_{t\to\infty} Z_t = \infty\} + P_z\{\lim_{t\to\infty} Z_t = 0\} = 1, \tag{2.9}$$

that $P_z(Z_t \to 0) = \exp(-z\rho)$ where ρ is the largest non-negative solution of $\psi(s) = 0$, and that $\rho = 0$ iff $D \leq 0$. It is routine to show, in the case $D > 0$, that the finite dimensional distributions of the CB-process conditioned on the event $Z_t \to 0$ coincides with those of the subcritical CB-process whose generating cf is $\psi(\rho + s)$. Theorems 4.5 and 4.6 of Li (1999) are examples of limit theorems for the supercritical case which can be read off from the subcritical results.

Since ψ is a concave function, it follows that the condition $\psi'(s) \geq 0$ for all s is necessary and sufficient for $\rho = \infty$, i.e., for $P_z(Z_t \to \infty) \equiv 1$. Differentiating (2.1) shows that the condition is equivalent to $\nu = 0$ and $a - \int_0^1 xQ(dx) \geq 0$. If D is finite, then the second of these can be expressed as $D - \int_0^\infty xQ(dx) \geq 0$.

3 Attaining the Origin

Until further notice there is no restriction on D, i.e. $D \leq \infty$. It is clear that the second integral in (2.1) is dominated by $Q((1,\infty)) < \infty$. Hence the determinants of the condition (2.8) are the Brownian component in (2.1) and the first integral there, denoted by $-K(s)$. The following result shows that the Brownian component dominates if it is present.

Lemma 3.1. $\lim_{s\to\infty} s^{-2}K(s) = 0.$

Proof. Observe that

$$0 < -(1-e^{-sx}-sx) = \int_0^{sx}\int_0^{y} e^{-z}\,dz\,dy \le s^2x^2/2 \quad (s>0),$$

and hence that the integrand in

$$s^{-2}K(s) = \int_0^1 \left[-s^{-2}(1-e^{-sx}-sx)\right] Q(dx)$$

is bounded above by x^2, and it converges to zero as $s \to \infty$. But $\int_0^1 x^2 Q(dx)$ is finite, and the assertion follows now from the dominated convergence theorem. □

The criterion (2.8) is always satisfied if the Brownian component is present. If $\nu = 0$, which we assume for the remainder of this section, then fulfillment of (2.8) depends on how fast $-\psi(s)$ grows to ∞. In particular, recall that (2.8) fails if $q_1 < \infty$. The following example provides a scale of comparison in terms of an index of regular variation, and it suggests that (2.8) is "usually" satisfied if $q_1 = \infty$.

Example 3.1. Let $\overline{Q}(x) = \int_0^x y^2 Q(dy)$, finite if $x > 0$. Feller's version of Karamata's Tauberian theorem (Bingham et al. (1987, Theorem 1.7.1′)) shows that

$$K''(s) = \int_0^1 e^{-sx}d\overline{Q}(x) = \Gamma(2-\rho)s^{-1+\rho}L(s) \quad (s\to\infty), \tag{3.1}$$

iff

$$\overline{Q}(x) \sim x^{1-\rho}L(1/x) \quad (x \to 0), \tag{3.2}$$

where $-\infty < \rho \le 1$ and L is slowly varying at infinity. In addition, $L(\infty) = 0$ when $\rho = 1$. Note that Q cannot be a Lévy measure if $\rho > 1$ or if $\rho = 1$ and $L(\infty) > 0$. A computation shows that $q_1 = \overline{Q}(1) + \int_0^1 y^{-2}\overline{Q}(y)\,dy$, and hence (3.2) implies that

$$q_1 = \begin{cases} \infty & \text{if } 0<\rho\le 1, \\ <\infty & \text{if } \rho < 0. \end{cases}$$

If $\rho = 0$, then the value of q_1 depends on the precise form of L. Noting that the second integral in (2.1) converges to $Q((1,\infty)) < \infty$ as $s \to \infty$, it follows that if $0 \le \rho \le 1$, then a two-fold integration of (3.1) yields

$$-\psi(s) = \text{const}\, s + s^{1+\rho}L_\rho(s), \tag{3.3}$$

where

$$L_\rho(x) \sim \frac{\gamma(2-\rho)}{\rho(1+\rho)}L(x) \quad (x\to\infty). \tag{3.4}$$

Clearly (2.8) is satisfied if $0 < \rho \leq 1$, that is, $q_1 = \infty$ (Case 3) and the origin is attainable. If $\rho = 0$ then

$$q_1 = \infty \text{ iff } \int_0^1 x^{-1}L(x^{-1})\,dx = \int_1^\infty x^{-1}L(x)\,dx = \infty.$$

This condition is satisfied if, for example, $L(x) \sim (\log x)^\delta$ and $\delta \geq -1$. But from (3.3) and (3.4), (2.8) is satisfied iff $\int_1^\infty (xL(x))^{-1}\,dx < \infty$, and this holds iff $\delta > 1$. In particular, the origin is *not* attainable if $-1 \leq \delta \leq 1$, even though $q_1 = \infty$.

In case (2.8) fails, the following result gives more detail about the nature of the DF $F_t(x|z) = P_z(Z_t \leq x)$ near its left-extremity $\ell_t = \sup\{x : F_t(x|1) = 0\}$. Since Z_t has a positive infinitely divisible law, its cf has a representation

$$\Psi(s,t) = \ell_t s + \int_0^\infty (1 - e^{-sx})n_t(dx),$$

where n_t is a Lévy measure, i.e., $\int_0^\infty (1 \wedge x)n_t(dx) < \infty$ for all $t \geq 0$. Recall that $\kappa = q_1 - a$ (see (2.7)) and note for the next result that we define $\kappa = \infty$ if $q_1 = \infty$.

Theorem 3.1. *If $D \leq \infty$ and (2.8) fails (and hence $\nu = 0$) then*

$$\ell_t = e^{-\kappa t}.$$

In particular, if $\nu = 0$ and $q_1 < \infty$ then $P_z(Z_t = z\ell_t) = \exp(-n_t(\mathbb{R}_+))$ and

$$n_t(\mathbb{R}_+) = \begin{cases} \frac{q_2}{\kappa}\left(1 - e^{-\kappa t}\right) & \text{if } \kappa \neq 0, \\ q_2 t & \text{if } \kappa = 0. \end{cases}$$

Proof. We give details only for $q_1 < \infty$ and $\kappa \geq 0$. Assume $\kappa > 0$ (which is the case if $D \leq 0$). Recalling that $\widetilde{Q}$ denotes the integral at (2.7), we let

$$\delta(s) = \frac{\widetilde{Q}(s)}{\kappa s(\kappa s - \widetilde{Q}(s))},$$

which is defined and positive for sufficiently large s. For such s, substitute $1/\psi(s) = -1/\kappa s - \delta(s)$ into (2.2) and integrate the result to obtain the identity

$$\Psi(s,t) - se^{-\kappa t} = se^{-\kappa t}\left[\exp\left(-\kappa \int_s^{\Psi(s,t)} \delta(u)\right) - 1\right]. \tag{3.5}$$

Since $\delta(s) = o(1/s)$, the term in square brackets vanishes as $s \to \infty$, and hence

$$\ell_t = \lim_{s\to\infty} \Psi(s,t)/s = e^{-\kappa t}.$$

The argument is easily adjusted to show that if $q_1 = \infty$ then $\ell_t = 0$.

To prove (ii) we need only evaluate $\widetilde{n}_t(\infty) = n_t(\mathbb{R}_+)$. But since $\Psi(s,t) = \ell_t s(1+o(1))$, the right-hand side of (3.5) is asymptotically proportional to

$$\kappa\ell_t s\int_{\ell_t s}^{s}\delta(u)\,du = \kappa\ell_t\int_{\ell_t}^{1} s^2\delta(ys)\,dy.$$

The assertion follows if $q_2 < \infty$ since

$$\kappa s^2\delta(ys) = \frac{\widetilde{Q}(s)}{\kappa y^2(1-\widetilde{Q}(ys)/s)} \to q_2/\kappa y^2.$$

If $\kappa = 0$ and $q_2 < \infty$, then the proof follows the same lines but now defining $\delta(s) = 1/\psi(s) - 1/q_2$, which leads to the identity

$$q_2 t = \Psi(s,t) - s + q_2\int_{s}^{\Psi(s,t)}\delta(u)\,du.$$

Finally, if $q_2 = \infty$ then altering some of the above details shows the limit of the right-hand side of (3.5) is infinite. □

The above results give a complete description of the behaviour of the law of Z_t at its left-extremity: $P_z(Z_t = 0) > 0$ iff (2.8) is satisfied; $\ell_t \equiv 0$ if (2.8) fails and $q_1 = \infty$; $\ell_t > 0$ iff $q_1 < \infty$, in which case $P_z(Z_t = z\ell_t) > 0$ iff $q_2 < \infty$.

We mention that if $\nu = 0$, $q_1 < \infty$ and $D \le 0$, then there exists a scaling function $0 < c_t \uparrow \infty$ such that $c_{t+\tau}/c_t \to 1/\ell_\tau$ and $c_t Z_t \to W$, a positive random variable with a non-degenerate law. If we further assume that $\mathcal{J} = \int_0^1 x(-\log x)Q(dx) < \infty$ then we can take $c_t = e^{\kappa t}$, the reciprocal of ℓ_t. See Grey (1974) for these facts. Pakes and Trajstman (1985) show when $\mathcal{J} < \infty$ and $c_t = e^{\kappa t}$ that the left-extremity of W is unity, and $P(W = 1) = q_2/\kappa$ when $q_2 < \infty$.

4 The Critical Case

Pakes and Trajstman (1985) gave some results for the critical subordinator case, $D = 0$, $q_1 < \infty$ *and* $\nu = 0$. It turns out that these restrictions are somewhat irrelevant. When $D = 0$ the representation (2.4) becomes

$$\psi(s) = -\nu s^2/2 + \int_0^\infty (1 - e^{-sx} - sx)Q(dx). \tag{4.1}$$

The key assumption we make is effectively one on the rate at which $Q((x,\infty))$ decreases as $x \to \infty$. Following Bingham (1976) we express it as follows: There is a positive function $b_t \uparrow \infty$ as $t \uparrow \infty$, and constants $\beta > 0$ and $0 < \delta \le 1$ such that

$$tb_t\psi(s/b_t) \to -\beta s^{1+\delta}/2\delta. \tag{4.2}$$

This is independent of the accessibility criterion (2.8).

A consequence of (4.2) is that b_t is regularly varying at infinity (RV) with index $1/\delta$. If $\delta < 1$ then the Brownian component in (4.1) makes no contribution. This also holds if $\delta = 1$ and $m_2 := \int_1^\infty x^2 Q(dx) = \infty$, since $b_t/t \to \infty$ in that case. Finally, if $m_2 < \infty$, then (4.2) holds with

$$\delta = 1, \quad \beta = \nu + m_2 \quad \text{and} \quad b_t \equiv t. \tag{4.3}$$

The most general limit result makes no reference to the accessibility of zero, and its form is as in Pakes and Trajstman (1985, p. 32). We reproduce it here.

Theorem 4.1. *If* (4.2) *holds then for each* $\varepsilon > 0$ *and* $x > 0$,

$$\lim_{t\to\infty} P_z\big(Z_t \le x b_t | Z_t > \varepsilon\big) = G(x),$$

where G is the distribution function whose Laplace–Stieltjes transform is

$$\widehat{G}(s) = 1 - \left[\frac{\beta s^\delta/2}{1+\beta s^\delta/2}\right]^{1/\delta}. \tag{4.4}$$

Proof outline. Bingham (1976) observes that (4.2) is equivalent to

$$b_t \Psi(s/b_t, t) \to \frac{s}{(1+\beta s^\delta/2)^{1/\delta}}. \tag{4.5}$$

With $G_t(x|z) = P_z(Z_t \le x b_t)$, the working of Pakes and Trajstman (1985, p. 33) holds unchanged in our more general context, yielding

$$b_t(1 - G_t(x|z)) \to z\Omega((x,\infty)) \tag{4.6}$$

where the finite measure Ω is specified by

$$\widetilde{\Omega}(s) = \frac{s}{(1+\beta s^\delta/2)^{1/\delta}}.$$

In addition

$$b_t P_z(Z_t > \varepsilon) \to z\Omega(\mathbb{R}_+) \quad \& \quad \Omega(\mathbb{R}_+) = (\beta/2)^{-1/\delta}. \tag{4.7}$$

The assertion follows upon setting $G(x) = \Omega((0,x])/\Omega(\mathbb{R}_+)$. □

Recall when (2.8) holds that T denotes the hitting time of zero. Let $\Psi(t) = \Psi(\infty, t)$, which is finite iff (2.8) holds. In this case it should be possible to set $\varepsilon = 0$ in Theorem 4.1. We state this as

Theorem 4.2. *If* (2.8) *and* (4.2) *hold, then for all* $x > 0$,

$$\lim_{t\to\infty} P_z(Z_t \le x b_t | T > t) = G(x).$$

Proof. Observe that

$$P_z(T > t) = 1 - e^{-z\Psi(t)} \sim z\Psi(t),$$

since $\Psi(t) \to 0$. In (2.2) let $s \to \infty$ and then $y = ub_t$ to get

$$\int_{b_t\Psi(t)}^{\infty} \frac{dy}{1 - tb_t\psi(y/b_t)} \equiv 1.$$

Letting $t \to \infty$ it is evident that $b_t\Psi(t) \to C$, say, and

$$\int_C^{\infty} y^{-1-\delta} dy = 2\delta/\beta, \quad \text{or} \quad C = (\beta/2)^{-1/\delta}.$$

We conclude that (c.f. (4.7))

$$b_t P_z(T > t) \to zC = z\Omega(\mathbb{R}_+).$$

□

If $m_2 < \infty$ then $G(x) = 1 - e^{-2x/\beta}$ where β is specified by (4.3) above, thus giving the CB-process version of (1.2). The structure of the limit law for $0 < \delta < 1$ is more complicated. It can be represented in terms of the following independent random variables. Let S_δ have the standard positive stable law with index δ; its Laplace–Stieltjes transform is $\exp(-s^\delta)$. Let Γ_a have the gamma(a) law; its Laplace–Stieltjes transform is $(1+s)^{-a}$. Finally, if X is a positive random variable such that $E(X^r) < \infty$ for some real r, then we define the length-bias operator of order r acting on (the law of) X *via*

$$P(\mathcal{L}_r X \le x) = \frac{1}{E(X^r)} \int_0^x y^r dP(X \le y).$$

See Pakes (1997a) for properties of $\mathcal{L}_r$.

Let χ be a random variable having the limiting distribution function G. Observe from (4.4) that

$$E(\chi) = (\beta/2)^{1/\delta}. \tag{4.8}$$

For the following discussion we will set the scale constant $\beta = 2$, thus defining the standard member of the parametric family defined by (4.4). Differentiating (4.4) yields

$$(1 + s^\delta)^{-1-1/\delta} = -\widehat{G}'(s). \tag{4.9}$$

The left-hand side is the Laplace–Stieltjes transform of the positive Linnik law $L(\delta, 1 + 1/\delta)$ (Pakes (1998a)). As shown there, this law can be represented by the product $\Gamma^{1/\delta}_{1+1/\delta} S_\delta$. The right-hand side of (4.9) is the Laplace–Stieltjes transform of $\mathcal{L}_1\chi$. This yields the 'in law' identity

$$\mathcal{L}_1\chi =_d \Gamma^{1/\delta}_{1+1/\delta} S_\delta.$$

Properties of $\mathcal{L}_r$ (Pakes (1997a, Lemma 2.1)) yield

$$\begin{aligned}\chi &=_d \mathcal{L}_{-1}\left(\Gamma^{1/\delta}_{1+1/\delta} S_\delta\right) \\ &=_d \left(\mathcal{L}_{-1}\Gamma^{1/\delta}_{1+1/\delta}\right)(\mathcal{L}_{-1}S_\delta) \\ &=_d (\mathcal{L}_{-1/\delta}\Gamma_1^{1+1/\delta})^{1/\delta}(\mathcal{L}_{-1}S_\delta) =_d \Gamma_1^{1/\delta}\mathcal{L}_{-1}S_\delta.\end{aligned}$$

Denoting Γ_1 by $\mathcal{E}$, which has the standard exponential law, we summarize the above discussion in the following theorem.

Theorem 4.3. *The limit law of Theorems* 4.1 *and* 4.2 *can be expressed as the product of independent factors*

$$\chi =_d \mathcal{E}^{1/\delta}\mathcal{L}_{-1}S_\delta.$$

The Laplace–Stieltjes transform of the second factor is

$$E\left[\exp(-s\mathcal{L}_{-1}S_\delta)\right] = \frac{\Gamma(1/\delta, s^\delta)}{\Gamma(1/\delta)}$$

where $\Gamma(a,x) = \int_x^\infty y^{a-1}e^{-y}\,dy$ *is an incomplete gamma function.*

The representation for χ reduces to $\mathcal{E}$ if $\delta = 1$. It is not possible to give a simple closed form expression for the density function of $\mathcal{L}_{-1}S_\delta$, unless $\delta = 1/2$, in which case the density function is $(4\sqrt{\pi x^5})^{-1}\exp(-1/4x)$, $x > 0$. Finally, Theorem 4.3 shows that χ has finite moments in a restricted range of orders: For $-1/\delta < r \leq 1+\delta$,

$$E(\chi^r) = \Gamma(1+r/\delta)\frac{E(S_\delta^{-1+r})}{E(S_\delta^{-1})} = \Gamma(1+r/\delta)\frac{\Gamma(1+(1-r)/\delta)}{\Gamma(1+1/\delta)\Gamma(2-r)}.$$

We look next at a sequence of corollaries of the fundamental Theorems 4.1 and 4.2. They involve conditioning on $Z_{t+\tau} > \varepsilon$ where τ is a function of t. If (2.8) holds and $\varepsilon = 0$ then this amounts to conditioning on deferring extinction until after some future time $t+\tau$. We examine three cases, but do not give the details when (2.8) holds. In addition to (4.5) and (4.7) we occasionally use the related assertion that

$$\lim_{t\to\infty, s_t\to s>0} b_t\Psi(s_t,t) = \Omega(\mathbb{R}_+). \tag{4.10}$$

The following preliminary result is a representation for the conditional distribution of Z_t/b_t.

Lemma 4.1.

$$P_z(Z_t \geq xb_t | Z_{t+\tau} > \varepsilon) = r(t,\tau)\int_x^\infty P_{yb_t}(Z_\tau > \varepsilon)d_yP_z(Z_t \leq yb_t | Z_t > \varepsilon),$$

where

$$r(t,\tau) = P_z(Z_t > \varepsilon)/P_z(Z_{t+\tau} > \varepsilon).$$

The next theorem shows that the limit law in Theorem 4.1 is preserved provided growth of τ is appropriately restricted.

Theorem 4.4. *Let the conditions of Theorem* 4.1 *hold, and suppose that* $\tau = o(t)$ *as* $t\to\infty$. *Then*

$$\lim_{t\to\infty} P_z(Z_t \leq xb_t | Z_{t+\tau} > \varepsilon) = G(x).$$

Proof. The proof is almost identical to that for Theorem 4.5 below. Here we have $r(t,\tau) \to 1$ and $P_{yb_t}(Z_\tau > \varepsilon) \to 1$ as $t \to \infty$. □

Our next result forces τ to grow in proportion to t.

Theorem 4.5. *Let the conditions of Theorem 4.1 hold, suppose that $\tau/t \to \alpha$, where $0 < \alpha < \infty$, and let $c = \alpha^{-1/\delta}\Omega(\mathbb{R})$. Then*

$$\lim_{t\to\infty} P_z\big(Z_t \le xb_t | Z_{t+\tau} > \varepsilon\big) = G(x,\alpha) \tag{4.11}$$

where $G(x,\alpha)$ is the distribution function whose Laplace–Stieltjes transform is

$$\widehat{G}(s,\alpha) = \frac{\widehat{G}(s) - \widehat{G}(s+c)}{1-\widehat{G}(c)}. \tag{4.12}$$

Proof. The regular variation of b_t and (4.6) yield $r(t,\tau) \to (1+\alpha)^{1/\delta} = 1/(1-\widehat{G}(c))$. In addition (4.10) shows that

$$\lim_{t\to\infty} E_{yb_t}\left(e^{-sZ_\tau}\right) = \lim_{t\to\infty} \exp\left[-y\frac{b_t}{b_\tau}\cdot b_\tau \Psi(s,\tau)\right] = e^{-cy}.$$

The limit is the Laplace–Stieltjes transform of the measure which assigns mass e^{-cy} to the origin and mass $1-e^{-cy}$ at ∞. Consequently

$$\lim_{t\to\infty} P_{yb_t}(Z_\tau > \varepsilon) = 1 - e^{-cy},$$

and (4.11) follows from Theorem 4.1 and Pratt's (1960) lemma with

$$1 - G(x,\alpha) = \frac{\int_x^\infty (1-e^{-cy})G(dy)}{1-\widehat{G}(c)}.$$ □

Using (4.8) and (4.9) we find the limit law has a finite first-order moment:

$$\int_0^\infty x dG(x,\alpha) = (\beta/2)^\delta \left[(1+\alpha)^{1/\delta} - \frac{\alpha^{1+1/\delta}}{1+\alpha}\right].$$

We have been unable to find a probabilistic representation of the limit law $G(x,\alpha)$ when $\delta < 1$, but see (4.16) for the case $\delta = 1$.

We now look at the case where τ grows unboundedly relative to t. The limit law we obtain is simply a scaled version of the $\mathrm{Li}(\delta, 1+1/\delta)$ law, mentioned above.

Theorem 4.6. *Let the conditions of Theorem 4.1 hold, and suppose that $\tau/t \to \infty$ as $t \to \infty$. Then*

$$\lim_{t\to\infty} P_z\big(Z_t \le xb_t | Z_{t+\tau} > \varepsilon\big) = G(x,\infty),$$

where

$$\widehat{G}(s,\infty) = (1+\beta s^\delta/2)^{-1-1/\delta}.$$

Proof. It follows from (4.7) that $(b_t/b_\tau)r(t,\tau) \to 1$ and

$$(b_\tau/b_t)\left(1 - E_{yb_t}\left(e^{-sZ_\tau}\right)\right) \to y(\beta/2)^{-1/\delta}.$$

The left-hand side is the cf of the measure which assigns mass $(b_\tau/b_t)P_{yb_t}(Z_\tau > \varepsilon)$ to (ε,∞), and hence this mass converges to the above limit. The limit assertion of the theorem again follows from Theorem 4.1, Lemma 4.1, and Pratt's lemma with

$$G(x,\infty) = (\beta/2)^{-1/\delta} \int_0^x yG(dy).$$

Hence $\widehat{G}(s,\infty) = -(\beta/2)^{-1/\delta}\widehat{G}'(s)$ and the remaining assertion follows from (4.4) and (4.8). □

Assigning a meaning to conditioning on $Z_{t+\tau} > \varepsilon$ when $\tau = \infty$ requires the following definitions. Let

$$\eta(s) = cs + \int_0^\infty (1 - e^{-sx})\Xi(dx)$$

be the cf of a subordinator (I_t), $(I_0 = 0)$. Here $c \geq 0$ and $\Xi(\cdot)$ is a Lévy measure,

$$\int_0^\infty (1 \wedge x)\Xi(dx) < \infty.$$

Intuitively, we define a continuous branching process with immigration, a CBI-process, as follows. A quantity dI_t of immigrant mass arrives during $(t, t+dt)$ and initiates a CB-process whose offspring cf is $\psi(s)$. Then Y_t is the total quantity at time t of the superposition of all the CB-processes initiated in this way, and they are independent. More precisely, (Y_t) is the Markov process whose transition law is specified by

$$E_z\left(e^{-sY_t}\right) = \exp\left[-z\Psi(s,t) - \int_0^t \eta(\Psi(s,u))du\right]. \tag{4.13}$$

See Kawazu and Watanabe (1971) for details of the construction.

The following theorem generalizes Theorem 4.1 of Li (2000) who assumes the origin is accessible. The proof is defered to §6.

Theorem 4.7. *If $D \leq 0$ then*

$$E_z\left(e^{-sZ_t}|Z_\infty > \varepsilon\right) := \lim_{\tau\to\infty} E_z\left(e^{-sZ_t}|Z_{t+\tau} > \varepsilon\right) = e^{-(z\Psi(s,t)+Dt)}\frac{\psi(\Psi(s,t))}{\psi(s)}. \tag{4.14}$$

The limit is the Laplace–Stieltjes transform of Y_t where (Y_t) is a CBI*-process whose immigration subordinator has the* cf $\eta(s) = \nu s + \int_0^\infty (1 - e^{-sx})xQ(dx)$.

In the particular case of the CB-diffusion process we have $Q(\cdot) \equiv 0$, and hence the immigration subordinator reduces to a steady accumulation of population mass at rate ν. We state a complement to Theorem 4.6 which is easily proved from (4.2), (4.5) and the following differentiated version of (2.2):

$$\Psi'(s,t) = \frac{\partial}{\partial s}\Psi(s,t) = \frac{\psi(\Psi(s,t))}{\psi(s)}, \quad s > 0. \tag{4.15}$$

Theorem 4.8. *Let the conditions of Theorem* 4.1 *hold. Then for each* $\varepsilon > 0$,

$$\lim_{t\to\infty} P(Z_t \le xb_t | Z_\infty > \varepsilon) = G(x,\infty).$$

If (2.8) is satisfied then the previous results hold with $\varepsilon = 0$, that is, conditionally on $T > t$. The proofs need minor adjustments along the lines of the proof of Theorem 4.2. We collect the outcomes of these adjustments into a single assertion as follows, and setting $G(x,0) = G(x)$.

Theorem 4.9. *Let* (2.8) *and the conditions of Theorem* 4.1 *hold, and suppose that* $\tau/t \to \alpha$ *where* $0 \le \alpha \le \infty$. *Then*

$$\lim_{t\to\infty} P_z(Z_t \le xb_t | T > t+\tau) = G(x,\alpha).$$

If $\delta = 1$ *then the limit law can be expressed as*

$$\chi(\alpha) = \frac{\beta}{2}\left(\mathcal{E} + \frac{\alpha}{\alpha+1}\mathcal{E}'\right) \tag{4.16}$$

where $\mathcal{E}$ *and* $\mathcal{E}'$ *are independent with a standard exponential law.*

If $\delta = 1$ then (4.4) becomes $\widehat{G}(s) = (1+\beta s/2)^{-1}$, the Laplace–Stieltjes transform of $\beta\mathcal{E}/2$. The representation (4.16) follows from (4.12). Its interpretation emanates from the observation that extinction may occur any time from the immediate to the remote future if $\alpha = 0$; that extinction is restricted to the intermediate future and beyond if $0 < \alpha < \infty$; and extinction is restricted to the remote future if $\alpha = \infty$. Thus Theorem 4.9 interpolates the known extreme cases $\alpha = 0$ and $\alpha = \infty$. This structure has been discussed for the simple BP by Pakes (1999).

The following result shows that the added α-dependent term $\alpha\mathcal{E}'/(1+\alpha)$ arises in its own right when we condition on extinction occurring *within* the intermediate future. Here, and in the sequel, we shall write F_t for $P_1(Z_t = 0)$ and $\overline{F}_t = 1 - F_t$.

Theorem 4.10. *Let* (2.8) *and the conditions of Theorem* 4.1 *hold, and suppose that* $\tau/t \to \alpha$ *where* $0 < \alpha \le \infty$. *Then*

$$\lim_{t\to\infty} P_z(Z_t \le xb_t | t < T \le t+\tau) = H(x,\alpha)$$

where $\widehat{H}(s,\alpha) = \widehat{G}(s+c)/\widehat{G}(c)$, *representing an exponential tilting of* G.

Proof. Observe that

$$P_z(t < T \leq t+\tau) = F^z_{t+\tau} - F^z_t = z(\overline{F}_t - \overline{F}_{t+\tau}) + o(\overline{F}_t)$$

whence, upon using results in the proof of Theorem 4.5,

$$\lim_{t\to\infty} b_t P_z(t < T \leq t+\tau) = z\widetilde{\Omega}(\infty)\big(1-(1+\alpha)^{-1/\delta}\big).$$

Next,

$$E_z\left(e^{-sZ_t}; t < T \leq t+\tau\right) = E_z\left(e^{-sZ_t}(F_\tau)^{Z_t}; Z_t > 0\right) = \exp\left(-z\Psi(s-\log F_\tau), t\right) - F^z_t.$$

By working as above and using (4.5) we obtain, since $b_t\overline{F}_\tau \to c$,

$$\lim_{t\to\infty} b_t E_z\left(e^{-sZ_t/b_t}; t < T \leq t+\tau\right) = z\left[\widetilde{\Omega}(\infty) - \widetilde{\Omega}(s+c)\right].$$

Some algebraic manipulation using $\widetilde{\Omega}(c) = \widetilde{\Omega}(\infty)(1+\alpha)^{-1/\delta}$ will complete the proof. □

5 Stationary Measures

We assume until further notice that $-\infty < D \leq \infty$. We call a nontrivial measure $m(dx)$ stationary if it is finite on bounded Borel sets, does not have an atom at the origin, and satisfies

$$m(A) = \int_0^\infty P_z(Z_t \in A)m(dz), \quad t > 0, \tag{5.1}$$

for any Borel set A not containing zero. In particular

$$m((0,x]) = \int_0^\infty P_z(0 < Z_t \leq x)m(dz) = \int_0^\infty \left[P_z(Z_t \leq x) - P_z(Z_t = 0)\right]m(dz). \tag{5.2}$$

Our first result shows under a weak finiteness condition (which Ogura (1970) includes as part of his definition) that there is no stationary measure when the origin is inaccessible.

Lemma 5.1. *Suppose* (2.8) *is* not *satisfied. Then there exists no stationary measure* $m(\cdot)$ *for which* $\widehat{m}(s) < \infty$ *for some* $s > 0$.

Proof. Suppose there exists $m(\cdot)$ and $s > 0$ for which $\widehat{m}(s) < \infty$. If (2.8) fails then the subtracted term in the integrand of (5.2) vanishes, so taking the Laplace–Stieltjes transform of (5.1) yields

$$\widehat{m}(s) = \int_0^\infty \exp[-z\Psi(s,t)]m(dz). \tag{5.3}$$

Recalling that $\rho := -\log P_1(\lim_{t\to\infty} Z_t \to 0)$ it follows from (2.9) that, as $t \uparrow \infty$, $\Psi(s,t)$ converges monotonically to ρ. So applying the monotone convergence theorem to (5.3) yields $\widehat{m}(s) \equiv \widehat{m}(\rho) < \infty$. Hence $m(dx) = \widehat{m}(\rho)\delta_0(dx)$, contradicting our definition of a stationary measure. □

Lemma 5.1 can be read as asserting that if there is a stationary measure with $\widehat{m}(s) < \infty$ for some $s > 0$, then the zero state is accessible. The finiteness assumption made in Lemma 5.1 is satisfied if $q_2 < \infty$ because, from Theorem 3.1, we have that

$$\infty > m((0,x]) \geq \int_0^\infty P_z(Z_t = \ell_t z) m(dz) = \widehat{m}(n_t(\mathbb{R}_+)).$$

Next we prove existence of a unique stationary measure, meaning unique up to multiplication by a constant. We use the fact that ρ is the largest solution of $\psi(s) = 0$.

Theorem 5.1. *If* (2.8) *holds then there exists a stationary measure whose Laplace–Stieltjes transform is*

$$\widehat{m}(s) = -\int_s^\infty dx/\psi(x), \quad s > 0, \tag{5.4}$$

finite if $s > \rho$. *Moreover it is the only stationary measure such that* $\widehat{m}(s) < \infty$ *for some* $s > \rho$.

Proof. Assume that $D \leq 0$, in which case $\rho = 0$. Small alterations to arguments used in Pakes (1997b, p. 283), shows that (5.4) defines a completely monotone function. Briefly, differentiating (5.4) yields

$$-\widehat{m}'(s) = \frac{1}{s} \cdot \left(\frac{-s}{\psi(s)} \right).$$

Let $q(x) = Q((x,\infty))$. Then, as observed by Pakes (1997b), $\int_0^\infty (1 \wedge x) q(x) dx < \infty$ and hence $q(\cdot)$ is the density function of the Lévy measure of a positive infinitely divisible law. In addition, $x^2 q(x) \to 0$ as $x \to \infty$. Thus integration by parts in (2.6) yields

$$-s/\psi(s) = \frac{1}{-D + \frac{1}{2}\nu s + \widetilde{q}(s)}$$

where

$$\widetilde{q}(s) = \int_0^\infty (1 - e^{-sx}) q(x)\, dx.$$

Hence, if (S_t) is the subordinator with cf $\frac{1}{2}\nu s + \widetilde{q}(s)$, we have

$$-s/\psi(s) = \int_0^\infty e^{Dt} E(\exp(-sS_t))\, dt = \int_0^\infty e^{-sx} d_x \int_0^\infty e^{Dt} P(S_t \leq x)\, dt$$

that is, the left-hand side is completely monotone. Since $1/s$ is completely monotone, so also is $-\widehat{m}'(s)$, whence $\widehat{m}(s)$ too is completely monotone. Denote its generating measure by $m(dx)$.

Next, (2.2) can be written as

$$\widehat{m}(\Psi(s,t)) - \widehat{m}(s) = t. \tag{5.5}$$

Observe that $\widehat{m}(\infty) = 0$ and $\Psi(t) := \Psi(\infty, t) = -\log F_t$. Let $s \to \infty$ in (5.5) and subtract the result from (5.5), obtaining

$$\widehat{m}(\Psi(s,t)) - \widehat{m}(-\log F_t) = \widehat{m}(s). \tag{5.6}$$

This is the Laplace–Stieltjes transform version of (5.2), thus showing that$m(dx)$ is indeed a stationary measure.

Finally, if $m(dx)$ is a stationary measure with $\widehat{m}(s') < \infty$, $s' > 0$, then (5.6) is valid for $s \geq s'$. But since $\Psi(s,t) < s$ for all $s > 0$, the range of validity of (5.6) can be extended to $s > 0$. Differentiation of (5.6) with respect to s and substituting from (4.15) yields the identity

$$\psi(s)\widehat{m}'(s) \equiv \psi(\Psi(s,t))\widehat{m}'(\Psi(s,t)).$$

Fixing s and allowing t to vary shows that the left-hand side is constant in$s$. Hence $\widehat{m}(s)$ is given by (5.4), apart from an arbitrary multiplicative constant.

Now let $D > 0$, and let $\overline{\psi}(s) = \psi(\rho + s) - \psi(\rho)$. Referring to (2.1), some manipulation shows that $\overline{\psi}$ is the cf of a subcritical CB-process $(\overline{Z}_t)$ whose defining triple is $(\overline{a}, \nu, \overline{Q})$ where

$$\overline{a} = a - 2\rho\nu - \int_{(0,1]} x(1 - e^{-\rho x})Q(dx) \quad \& \quad \overline{Q}(dx) = e^{-\rho x}Q(dx).$$

Some manipulation with the backward equation (2.2) shows that the cf's of the two CB-processes are related through $\Psi(s,t) = \rho + \overline{\Psi}(s - \rho, t)$. It follows that if $F_t(dx|z)$ (resp. $\overline{F}_t(dx|z)$) denotes the transition kernel of (Z_t) (resp. $\overline{Z}_t$) then

$$e^{\rho z}F_t(dx|z) = e^{\rho x}\overline{F}_t(dx|z).$$

The first part of the proof establishes the existence of the stationary measure$\overline{m}(dx)$ of $(\overline{Z}_t)$, and it follows that the relation

$$\overline{m}(dx) = \int_0^\infty \overline{F}_t(dx|z)\overline{m}(dz)$$

is equivalent to (5.1) with $m(dx) = e^{\rho x}\overline{m}(dx)$. □

Remark 5.1. If $D \leq 0$ the proof that $\widehat{m}(s)$ is completely monotone shows that $m(dx)$ has the representation

$$m(dx) = \left(\frac{1}{x} \int_0^\infty e^{Dt} P(S_t \leq x)dt \right) dx, \tag{5.7}$$

and hence that it is absolutely continuous. A simple modification gives a similar result for the case $D > 0$.

Letting $s \to \infty$ in (5.5) gives the useful representation

$$\widehat{m}(-\log F_t) = t. \tag{5.8}$$

For the remainder of this section we will assume $D = 0$ and show how the stationary measure is related to a conditional limit theorem for the population size. We will need the following lemma in which $\mu(t)$ is the inverse function of $\widehat{m}(\cdot)$. This inverse is defined for $t > 0$, $\mu(0+) = \infty$, and $\mu(t) \downarrow 0$ as $t \uparrow \infty$.

Lemma 5.2. *Suppose that $D = 0$ and (2.8) both hold. If $t' > 0$ is fixed, then*

$$\lim_{t\to\infty} \frac{\psi(\mu(t+t'))}{\psi(\mu(t))} = 1.$$

Proof. Let $s' = \mu(t)$ and $s'' = \mu(t+t')$. Then

$$1 - \frac{\psi(\mu(t+t'))}{\psi(\mu(t))} = \frac{\psi(s') - \psi(s'')}{\psi(s')} = \frac{(s'-s'')(-\psi'(\bar{s}))}{-\psi(s')},$$

where $s'' < \bar{s} < s'$, and $\psi'(\bar{s}) \to 0$ $(t \to \infty)$ since $D = 0$. But

$$t' = \widehat{m}(s'') - \widehat{m}(s') = \int_{s''}^{s'} \frac{dv}{-\psi(v)} > \frac{s'-s''}{-\psi(s')} > 0. \qquad \square$$

Since $\ell_t \equiv 0$ if (2.8) holds, the Lévy representation for the cf of Z_t is $\Psi(s,t) = \widetilde{n}_t(s)$. Moreover, since $F_t > 0$ the Lévy measure $n_t(\cdot)$ is totally finite, i.e., Z_t has a compound Poisson law for each t. The following result is a ratio limit theorem for $n_t(\cdot)$. In the remainder of this section, c will denote an arbitrary constant not related to its meaning in §4. Recall that $\Psi(t) := \Psi(\infty, t)$.

Theorem 5.2. *Let $D = 0$ and (2.8) both hold. For each $c > 0$ and Borel set $A \in \mathbb{R}_+$,*

$$\lim_{t\to\infty} \frac{n_t(A)}{\widehat{n}_t(c)} = \frac{m(A)}{\widehat{m}(c)}.$$

In particular, if A and B are Borel subsets of $\mathbb{R}_+$ and carrying finite stationary measure, then

$$\lim_{t\to\infty} \frac{n_t(A)}{n_t(B)} = \frac{m(A)}{m(B)}.$$

Proof. For each $c > 0$ we see that

$$\widehat{M}_c(s,t) := \frac{\widehat{n}_t(c+s)}{\widehat{n}_t(c)} = \frac{\Psi(t) - \Psi(c+s,t)}{\Psi(t) - \Psi(c,t)}$$

is the Laplace–Stieltjes transform of the distribution function

$$M_c(x,t) = (\widehat{n}_t(c))^{-1} \int_0^x e^{-cy} n_t(dy).$$

Rewrite (5.5) as

$$\Psi(s,t) = \mu(t + \widehat{m}(s)). \tag{5.9}$$

We infer from (5.4) that $\mu'(t) = \psi(\mu(t))$ and, since $\mu(\infty) = 0$, we have

$$\mu(t) = -\int_t^\infty \psi(\mu(y))\, dy.$$

It follows now from (5.9) that

$$\widehat{M}_c(s,t) = \frac{\int_t^{t+\widehat{m}(c+s)} \psi(\mu(y))\,dy}{\int_t^{t+\widehat{m}(c)} \psi(\mu(y))\,dy} = \frac{\widehat{m}(c+s)}{\widehat{m}(c)} \cdot \frac{\psi(\mu(y'))}{\psi(\mu(y''))},$$

where $t < y' < t+\widehat{m}(c+s)$ and $t < y'' < t+\widehat{m}(c)$. It is clear from Lemma 5.2 that the second factor on the right-hand side tends to unity as $t \to \infty$. Hence the continuity theorem for Laplace–Stieltjes transforms yields

$$\frac{1}{\widehat{n}_t(c)} \int_A e^{-cx} n_t(dx) \to \frac{1}{\widehat{m}(c)} \int_A e^{-cx} m(dx),$$

and the assertion follows. □

It is worth isolating the following corollary of the proof of Theorem 5.2.

Lemma 5.3. *Let $0 < s', s'' \le \infty$. Under the conditions of Theorem 5.2,*

$$\lim_{t\to\infty} \frac{\Psi(t) - \Psi(s',t)}{\Psi(t) - \Psi(s'',t)} = \frac{\widehat{m}(s')}{\widehat{m}(s'')}.$$

In particular, if $t' \ge 0$ and $t'' > 0$ then

$$\lim_{t\to\infty} \frac{\Psi(t) - \Psi(-\log F_{t'},t)}{\Psi(t) - \Psi(-\log F_{t''},t)} = t'/t''.$$

The next result is a conditional limit theorem allowing for extinction occurring in the near-future, to be compared with Theorem 2.5 in Pakes (1999) for the simple BP. We let A denote a Borel set which can be represented as a countable union of bounded and disjoint intervals, $A = \bigcup_{n\ge 1} (t'_n, t''_n]$, where $t'_n < t''_n < t'_{n+1}$.

Theorem 5.3. *Let $D = 0$ and (2.8) both hold. Then*

$$\lim_{t\to\infty} P_z(Z_t \le x | T \in t+A) = \begin{cases} \dfrac{1}{|A|} \displaystyle\int_0^x P_y(T \in A) m(dy) & \text{if } |A| < \infty, \\ 0 & \text{if } |A| = \infty. \end{cases} \tag{5.10}$$

The limit law is non-defective if $|A| < \infty$.

Proof. We first prove the result for $A = (t', t'']$, $0 \le t' < t'' < \infty$. Let $s' = -\log F_{t'}$ and $s'' = -\log F_{t''}$. It is clear that

$$E_z\left(e^{-sZ_t} | t+t' < T \le t+t''\right) \sim \frac{\Psi(s+s',t) - \Psi(s+s'',t)}{\Psi(s',t) - \Psi(s'',t)}. \tag{5.11}$$

Dividing the numerator and denominator by $\Psi(t) - \Psi(s'',t)$ and regrouping terms so that Lemma 5.3 can be used, it is easily seen that the right-hand side of (5.11) converges as

$t \to \infty$ to $\left[\widehat{m}(s+s'') - \widehat{m}(s+s')\right]/(t''-t')$ which, for the present choice of A, is the Laplace–Stieltjes transform of the right-hand side of (5.10). To see that the limit is a non-defective distribution function, first observe that

$$P_y(T \in A) = \int_{0+}^{\infty} F^x_{|A|} F_{t'}(dx|y)$$

which, along with (5.1) and (5.8), entail that

$$\int_0^{\infty} P_y(T \in A) m(dy) = \int_{0+}^{\infty} F^x_{|A|} m(dx) = \widehat{m}(-\log F_{|A|}) = |A|.$$

Now let A be the countable union of intervals as above, and define s'_n and s''_n analagously to s' and s''. Observe that

$$E_z(e^{-sZ_t}; T \in t+A) \sim z \sum_{n\geq 1} \left[\Psi(s+s'_n,t) - \Psi(s+s''_n,t)\right].$$

Suppose first that $|A| = \infty$. If $s > 0$ then the sum is bounded above by

$$\sum_{n\geq 1} \left[\Psi(s+s''_{n+1},t) - \Psi(s+s''_n,t)\right] \leq \Psi(\infty,t) - \Psi(s+s''_1,t).$$

Lemma 5.3 shows that the upper bound is asymptotically proportional to

$$\frac{\widehat{m}(s) - \widehat{m}(s+s''_1)}{\widehat{m}(1)} \left(\Psi(t) - \Psi(1,t)\right).$$

Similarly, by Fatou's lemma,

$$\begin{aligned} \liminf_{t\to\infty} \frac{P_z(T \in t+A)}{z(\Psi(t) - \Psi(1,t))} &\geq \sum_{n\geq 1} \frac{\widehat{m}(s''_n) - \widehat{m}(s'_n)}{\widehat{m}(1)} \\ &= \sum_{n\geq 1} \frac{t''_n - t'_n}{\widehat{m}(1)} = |A|/\widehat{m}(1) = \infty. \end{aligned}$$

The assertion for $|A| = \infty$ follows from these estimates.

If $|A| < \infty$ then the bounding technique used above can be adjusted to validate use of the dominated convergence theorem on the sum

$$\frac{\sum_{n\geq 1} \left[\Psi(s+s''_{n+1},t) - \Psi(s+s''_n,t)\right]}{\Psi(t) - \Psi(1,t)},$$

now for $s \geq 0$. The details are omitted as they are similar to those for Theorem 6.7 below. □

Example 5.1. If $0 < \delta < 1$, $a = \nu = 0$ and

$$Q(dx) = \frac{\beta}{2} \cdot \frac{1+\delta}{\Gamma(1-\delta)} x^{-2-\delta} dx$$

then

$$\psi(s) = -\frac{\beta}{2\delta} s^{1+\delta}.$$

We obtain this form with $\delta = 1$ if $a = 0$, $Q(\cdot) \equiv 0$ and $\beta = \nu > 0$, i.e. (Z_t) is the CB-diffusion.

Integration yields $\widehat{m}(s) = (2/\beta)s^{-\delta}$ and inversion gives

$$m(dx) = \frac{2}{\beta\Gamma(\delta)} x^{\delta-1} dx.$$

In particular, the stationary measure for the CB-diffusion is proportional to Lebesgue measure.

Inverting (5.8) yields $F_t = \exp(-(2/\beta t)^{1/\delta})$. For fixed $\tau > 0$, Theorem 5.3 yields

$$\lim_{t\to\infty} P_z\big(Z_t \leq x | t < T \leq t+\tau\big) = \frac{2}{\beta\tau\Gamma(\delta)} \int_0^x y^{\delta-1} e^{-y(2/\beta\tau)^{1/\delta}} dy,$$

the distribution function of a gamma law.

6 The Subcritical Case

Consider now the subcritical case $D < 0$. Following Pakes and Trajstman (1985, p. 34), define

$$h(s) = \exp\left[D \int_1^s \frac{du}{\psi(u)}\right]. \tag{6.1}$$

The next lemma develops several properties of this function. We use the abbreviation SV for slow variation at infinity.

Lemma 6.1. *The function h satisfies:*

(a) $h(0+) = 0$ *and h is concave increasing;*

(b) $h(\infty) < \infty$ *iff* (2.8) *holds;*

(c) $h(s) = s/L(1/s)$ *where L is SV; and*

(d) $L(\infty) > 0$ *iff*

$$\int_0^1 (u^{-1} - D/\psi(u))\, du < \infty,$$

i.e., iff

$$\int_1^\infty (x \log x) Q(dx) < \infty. \tag{6.2}$$

Proof. Assertions (a) and (b) are obvious. The representation in (c) follows because

$$h(s)/s = \exp\left[-\int_1^s (u^{-1} - D/\psi(u))du\right]. \tag{6.3}$$

The substitution $x = 1/u$ shows that the reciprocal of the right-hand side can be expressed as $L(1/s)$ where

$$L(z) = \exp\left[\int_1^z \varepsilon(x)dx/x\right] \quad \& \quad \varepsilon(x) = 1 - D/x\psi(1/x).$$

Since ψ is concave and decreasing, we see that $\varepsilon(x) \downarrow 0$ as $x \uparrow \infty$. Thus L has the characteristic representation of a SV function; see Bingham et al. (1987, p. 15).

The integrand in (6.3) is positive, and hence $L(\infty) > 0$ iff

$$\int_0^1 (1/u - D/\psi(u))\, du < \infty$$

or, since $\psi(u) \sim Du$ near the origin, iff

$$-\int_0^1 (\psi(s) - Ds)s^{-2}\, ds < \infty.$$

The Brownian component does not affect this condition, and hence it holds iff

$$-\int_0^1 s^{-2} \int_1^\infty (1 - e^{-sx} - sx)Q(dx)\, ds < \infty.$$

Assertion (d) follows by reversing the order of integration and observing that the resulting inner integral grows, as a function of x, like $x \log x$. □

Let $g(v)$ denote the inverse function of $h(s)$. Its properties can be inferred from Lemma 6.1, and those we need are listed as follows.

Lemma 6.2. *The function g is defined on $[0, h(\infty))$ and:*

(a) $g(0) = 0$ *and g is increasing;*

(b) *g is defined in $\mathbb{R}_+$ iff* (2.8) *fails;*

(c) $g(v) = v/L^*(1/v)$, *where L^* is the* SV *conjugate of L; and*

(d) $L^*(\infty) > 0$ *iff* (6.2) *holds.*

The following developments have obvious similarities to Pakes and Trajstman (1985, p. 34). They are rendered more direct by expressing $\Psi(s,t)$ in terms of h and g as follows. The backward equation (2.2) can be written as $\log h(\Psi(s,t)) - \log h(s) = Dt$, i.e.

$$\Psi(s,t) = g(e^{Dt}h(s)). \tag{6.4}$$

Lemma 6.2 implies that

$$\lim_{t\to\infty}\frac{\Psi(s,t)}{\Psi(1,t)} = h(s). \tag{6.5}$$

This is the core of the following ratio limit theorem (c.f. Theorem 5.2).

Theorem 6.1. *The function h is the cf of a positive infinitely divisible law:*

$$h(s) = \int_0^\infty (1-e^{-sx})\lambda(dx), \tag{6.6}$$

where λ is a Lévy measure. In addition, for Borel subsets A and B with $\lambda(B) < \infty$,

$$\lim_{t\to\infty}\frac{n_t(A)}{\Psi(1,t)} = \lambda(A) \quad \& \quad \lim_{t\to\infty}\frac{n_t(A)}{n_t(B)} = \frac{\lambda(A)}{\lambda(B)}. \tag{6.7}$$

The measure λ is finite iff (2.8) holds.

Proof. It follows from (6.4) and the continuity theorem for cf's (Appendix in Seneta and Vere-Jones (1968)) that h is a cf with spectral measure λ and a drift parameter q, say, i.e. $h(s) = qs + \widetilde{\lambda}(s)$.

We show in all cases that $q = 0$. This is obvious when (2.8) holds, as then $h(\infty) < \infty$. Suppose (2.8) fails. If also $q_1 = \infty$ then $\ell_t \equiv 0$ and the continuity theorem for cf's shows again that $q = 0$. If $q_1 < \infty$ then $\kappa > -D$. But (6.4) shows that $\Psi(1,t) \sim h(1)e^{Dt}/L^*(e^{-Dt})$ and it follows from (3.5) that

$$q = \lim_{t\to\infty} \ell_t/\Psi(1,t) = 0.$$

This establishes (6.6), and (6.7) follows from the continuity theorem for cf's. The final assertion follows from (6.1) and (6.6). □

The representation (6.4) is equivalent to the functional equation

$$h(\Psi(s,t)) = e^{Dt}h(s). \tag{6.8}$$

Suppose (2.8) holds, and recall that $F_t = \exp(-\Psi(t))$. Then $h(\infty) = \lambda(\mathbb{R}_+) < \infty$, and letting $s \to \infty$ in (6.8) gives the representation

$$h(-\log F_t) = h(\infty)e^{Dt}. \tag{6.9}$$

Writing $h(s) = \lambda(\mathbb{R}_+) - \widehat{\lambda}(s)$, (6.8) and (6.9) yield

$$\widehat{\lambda}(\Psi(s,t)) - \widehat{\lambda}(\Psi(t)) = e^{Dt}\widehat{\lambda}(s)$$

or

$$\int_0^\infty P_z(Z_t \in A)\lambda(dz) = e^{Dt}\lambda(A), \tag{6.10}$$

whenever $A \subset (0,\infty)$. *Thus the finite measure* λ *is* $(-D)$*-invariant for the transition semi-group of* (Z_t). This provides an alternative approach to the final assertion of Theorem 4.3 in Li (2000).

Now suppose (2.8) does not hold - the zero state is inaccessible. If a $(-D)$-invariant measure M exists, then it satisfies (6.10). If, in addition, $\widehat{M}(s) < \infty$ for some $s > 0$ then

$$\int_0^\infty e^{-z\Psi(s,t)} M(dz) = e^{Dt}\widehat{M}(s).$$

The right-hand side tends to 0 as $t \to \infty$, but the left-hand side tends to $M((0,\infty))$, by virtue of the monotone convergence theorem. Hence there exists no non-trivial $(-D)$-invariant measure if (2.8) fails.

The following theorems show that λ determines limit laws for conditional limit theorems whether or not (2.8) holds. Let $\mathcal{L}(x) = \lambda((x,\infty))$, which is finite if $x > 0$ because λ is a Lévy measure. The following result contains the fundamental conditional limit theorem for (Z_t) when $D < 0$.

Theorem 6.2. *Let* $D < 0$. *For any Borel set* $A \subset \mathbb{R}_+$,

$$\lim_{t\to\infty} P_z(Z_t \in A)/\Psi(1,t) = z\lambda(A).$$

In particular, if $\varepsilon > 0$ *then*

$$\lim_{t\to\infty} P_z(Z_t \le x | Z_t > \varepsilon) = \begin{cases} 1 - \mathcal{L}(x)/\mathcal{L}(\varepsilon) & \text{if } x \ge \varepsilon, \\ 0 & \text{if } x < \varepsilon. \end{cases}$$

Proof. The cf

$$\frac{\int_0^\infty (1-e^{-sx})F_t(dx|z)}{\Psi(1,t)} = \frac{1-e^{-z\Psi(s,t)}}{\Psi(1,t)} \sim z\Psi(s,t)/\Psi(1,t).$$

Referring to (6.5), the first assertion follows from (6.6) and the continuity theorem for cf's. The second assertion follows from the first. □

A subcritical analogue of Theorem 4.4 can be obtained from the following version of Lemma 4.1:

$$P_z(Z_t \le x | Z_{t+\tau} > \varepsilon) = \frac{\int_0^x P_y(Z_\tau > \varepsilon)F_t(dy|z)}{P_z(Z_{t+\tau} > \varepsilon)}.$$

If $\tau > 0$ is fixed, then Lemma 6.2(c), Theorem 6.2 and (6.4) imply that

$$P_z(Z_{t+\tau} > \varepsilon)/\Psi(1,t) \to ze^{D\tau}\lambda((\varepsilon,\infty)).$$

These results, together with Theorem 6.2 expressed in the form

$$F_t(dy|z))/\Psi(1,t) \to z\lambda(dy),$$

imply the next result, after noting that the honesty of the limit law follows from the invariance equation (6.10).

Theorem 6.3. *If $D < 0$ and $\varepsilon, \tau > 0$ are fixed, then*

$$J_\varepsilon(x,\tau) := \lim_{t\to\infty} P_z(Z_t \le x | Z_{t+\tau} > \varepsilon) = e^{-D\tau} \frac{\int_0^x P_y(Z_\tau > \varepsilon)\lambda(dy)}{\lambda((\varepsilon,\infty))}.$$

The next result shows that, in contrast to Theorems 4.5 and 4.6, the rate at which $\tau \to \infty$ has no influence on the form of limit law.

Theorem 6.4. *If $D < 0$, $\varepsilon > 0$, and $\tau \to \infty$ as $t \to \infty$, then*

$$J(x) := \lim_{t\to\infty} P_z(Z_t \le x | Z_{t+\tau} > \varepsilon) = \begin{cases} \dfrac{\int_0^x y\lambda(dy)}{\int_0^\infty y\lambda(dy)} & \text{if (6.2) holds,} \\ 0 & \text{if (6.2) fails.} \end{cases}$$

Proof. If (6.2) holds then the slowly varying functions in Lemmas 6.1(d) and 6.2(c,d) satisfy $L(\infty)L^*(\infty) = 1$. With $s_\tau = -\log F_\tau(\varepsilon)$, (6.4) yields

$$\Psi(s+s_\tau,t) - \Psi(s,t) = [h(s+s_\tau) - h(s)] \left(e^{Dt}/L^*(\infty)\right)(1+o(1))$$

and

$$\Psi(s_\tau,t) \sim h(s_\tau)e^{Dt}/L^*(\infty) \sim s_\tau e^{Dt}/L(\infty)L^*(\infty) = s_\tau e^{Dt},$$

both as $t \to \infty$. Hence

$$E_z(e^{-sZ_t}|Z_{t+\tau} > \varepsilon) \to h'(s)/L^*(\infty).$$

But $L^*(\infty) = 1/L(\infty) = h'(0)$, so $h'(s)/h'(0)$ is the Laplace–Stieltjes transform of the non-defective limiting distribution function in the assertion. If (6.2) fails, we simply note that $h'(0) = \infty$, omitting further detail. □

Proof of Theorem 4.7. We have now established most of the preliminary results we need. All that remains is an extension of (6.5) to the case $D = 0$ for which, in consonance with (6.1), we define $h(s) \equiv 1$. To obtain this, recall that $\mu(t)$ is the inverse of $\widehat{m}(s)$, that $\mu(\infty) = 0$, and $\mu'(t) = \psi(\mu(t))$. Defining $N(x) = \mu(\log x)$, we obtain

$$xN'(x)/N(x) = \mu'(\log x)/\mu(\log x) = \psi(N(x))/N(x) \to 0 \text{ as } x \to \infty,$$

showing that $N(x)$ is SV. Hence

$$\lim_{t\to\infty} \frac{\Psi(s,t)}{\Psi(1,t)} = \lim_{t\to\infty} \frac{N(e^{t+\widehat{m}(s)})}{N(e^{t+\widehat{m}(1)})} = 1,$$

that is, (6.5) holds if $D \leq 0$. Assuming this, we obtain

$$\lim_{\tau\to\infty} \frac{1-\exp(-z\Psi(s,\tau))}{\Psi(1,\tau)} = z\frac{h(s)}{h(1)},$$

and the continuity theorem for cf's yields

$$\lim_{\tau\to\infty} \frac{P_z(Z_\tau > \varepsilon)}{\Psi(1,\tau)} = z\frac{\lambda((\varepsilon,\infty])}{h(1)} \quad (\varepsilon > 0). \tag{6.11}$$

For positive numbers u, τ and ε,

$$z\Psi(1,t) > 1-e^{-z\Psi(1,\tau)} = E_z\left(1-e^{-Z_\tau}\right) > \left(1-e^{-\varepsilon}\right)P_z(Z_\tau > \varepsilon).$$

It follows from (6.11) and the dominated convergence theorem that, if $s \geq 0$, then

$$\lim_{\tau\to\infty} \frac{E_z\left[e^{-sZ_t}P_{Z_t}(Z\tau > \varepsilon)\right]}{\Psi(1,\tau)} = \frac{\lambda((\varepsilon,\infty])}{h(1)}E_z\left(Z_t e^{-sZ_t}\right).$$

Consequently

$$\begin{aligned}\lim_{\tau\to\infty} E_z(e^{-sZ_t}|Z_{t+\tau} > \varepsilon) &= \lim_{\tau\to\infty} \frac{E_z\left[e^{-sZ_t}P_{Z_t}(Z_\tau > \varepsilon)\right]}{E_z[P_{Z_t}(Z_\tau > \varepsilon)]} \\ &= \frac{E_z\left[e^{-sZ_t}Z_t\right]}{E_z[Z_t]} = \Psi'(s,t)e^{-z\Psi(s,t)}/e^{Dt}.\end{aligned}$$

The second equality in (4.14) now follows from (4.15). Checking that the last term in (4.14) has the form of (4.13) is accomplished by setting $z = 0$, and then using the representation (2.6) and setting $\eta(s) = D - \psi'(s)$ in (4.15). The integral in (4.13) is evaluated using the change of variable $y = \Psi(s,u)$, and in conjunction with the backward equation this gives $du = dy/\psi(y)$. □

The next result identifies the limit law J in Theorem 6.4 as the limiting-stationary law of the CBI-process of Theorem 4.7.

Theorem 6.5. *If $D < 0$, then*

$$\lim_{t\to\infty} P(Z_t \leq x | Z_\infty > \varepsilon) = J(x).$$

Proof. Since $\psi(s) \sim Ds$ $(s \to 0)$, (6.4) together with Lemmas 6.1 and 6.2 show that the right-hand side of (4.14) converges to $Dh(s)/\psi(s)L^*(\infty)$. This is identically zero if (6.2) fails. If (6.2) holds, then the definition (6.1) shows that the limit equals $h'(s)/L^*(\infty) = h'(s)/h'(0)$. This is the Laplace–Stieltjes transform of $J(x)$, as shown in the proof of Theorem 6.4. □

Remark. *In Theorem* 6.3 *the limit law* $J_\varepsilon(x,\tau) \to J(x)$ *as* $\tau \to \infty$.

The above proofs hold if the zero state is accessible and $\varepsilon = 0$. The conditioning event in Theorem 6.2 is replaced by $\{T > t\}$, and in Theorems 6.3 and 6.4 by $\{T > t+\tau\}$. Since $\mathcal{L}(0) = \lambda(\mathbb{R}_+) < \infty$ if (2.8) holds, the fundamental limit assertion of Theorem 6.2 takes the following form.

Theorem 6.6. *If* $D < 0$ *and* (2.8) *both hold, then*

$$\lim_{t\to\infty} P_z(Z_t \le x | T > t) = \lambda((0,x])/\lambda(\mathbb{R}_+).$$

The next result is a formal generalization of Theorem 6.6, similar to Theorem 5.3, and analogous to Theorem 2.1 in Pakes (1999) for the simple branching process. Let A be a countable union of intervals, as described before Theorem 5.3, and let $\mathcal{E}_D$ be a random variable having the exponential law $P(\mathcal{E}_D > t) = e^{Dt}$.

Theorem 6.7. *Let* $D < 0$ *and* (2.8) *both hold. Then*

$$\lim_{t\to\infty} P_z(T \in t + A)/\Psi(1,t) = zP(\mathcal{E}_D \in A)$$

and

$$\lim_{t\to\infty} P_z(Z_t \le x | T \in t + A) = \frac{\int_0^x P_z(T \in A)\lambda(dz)}{\lambda(\mathbb{R}_+)P(\mathcal{E}_D \in A)}.$$

Proof. Define s'_n and s''_n as in the proof of Theorem 5.3. For $s \ge 0$ we have as $t \to \infty$ that

$$(\Psi(1,t))^{-1}E_z(e^{-sZ_t}; T \in t + A) \sim (z/\Psi(1,t)) \sum_{n\ge 1} \left[\Psi(s + s'_n, t) - \Psi(s + s''_n, t)\right].$$

The nth summand is bounded above by

$$\Delta_n(t) = \Psi(s + s'_n, t) - \Psi(s + s'_{n+1}, t).$$

Added over the range $n \ge 1$, these bounds total $\Psi(s + s'_1, t) - \Psi(s + s'_\infty, t)$, where $s'_\infty = \lim_{n\to\infty} s'_n \le \infty$. It follows from (6.5) that

$$\sum_{n\ge 1} \lim_{t\to\infty} \Delta_n(t) = \lim_{t\to\infty} \sum_{n\ge 1} \Delta_n(t) = h(s + s'_1) - h(s + s'_\infty) < \infty.$$

This shows that Pratt's (1960) lemma is applicable, giving

$$\lim_{t\to\infty} (\Psi(1,t))^{-1}E_z(e^{-sZ_t}; T \in t + A) = z \sum_{n\ge 1} \left[h(s + s'_n) - h(s + s''_n)\right] \quad (s \ge 0), \tag{6.12}$$

and if $s > 0$ then the right-hand side equals

$$z \sum_{n\ge 1} \int_0^\infty e^{-sz} \left(F^z_{t''_n} - F^z_{t'_n}\right) \lambda(dz) = \int_0^\infty e^{-sz} P_z(T \in A)\lambda(dz).$$

If $s = 0$ we see from (6.9) that the limit (6.12) equals

$$zh(\infty) \sum_{n\ge 1} \left(e^{Dt'_n} - e^{Dt''_n}\right) = z\lambda(\mathbb{R}_+)P(\mathcal{E}_D \in A),$$

and the assertion follows. □

Next we determine some details about the structure of the limit laws in Theorems 6.2 and 6.6. Recall that $\mathcal{L}(x) = \lambda((x,\infty))$. Expressing (6.6) as a repeated integral gives

$$\omega(s) := \int_0^\infty e^{-sx}\mathcal{L}(x)\,dx = h(s)/s. \tag{6.13}$$

Observe that $\int_0^1 \mathcal{L}(x)\,dx = \int_0^1 x\lambda(dx) < \infty$. The properties asserted in the following lemma reflect analagous properties of the subcritical MBP, for which see Asmussen and Hering (1983).

Lemma 6.3.

(a) $\int_0^x \mathcal{L}(u)du \sim 1/L(x)$ *as* $x \to \infty$*; and*

(b) *The limit laws in Theorems* 6.2 *and* 6.6 *have finite means iff* (6.2) *holds.*

(c) *If* (2.8) *holds and* F *denotes the limit* DF *of Theorem* 6.6*, then*

$$\int_0^\infty e^{-sx}F(dx) = 1 - \exp\left[-D\int_s^\infty du/\psi(u)\right]. \tag{6.14}$$

In particular, $F(0) = 0$*;*

$$\mu = \int_0^\infty xF(dx) = \exp\left[\int_0^1 \left(u^{-1} - D/\psi(u)\right)du - \int_1^\infty (D/\psi(u))du\right]; \tag{6.15}$$

and

$$\lim_{t\to\infty} e^{-Dt}P_z(T > t) = z/\mu.$$

Proof. Lemma 6.1(c) implies that

$$\int_0^\infty e^{-sx}d_x\int_0^x \mathcal{L}(y)dy = 1/L(1/s).$$

The first assertion follows from Karamata's Tauberian theorem; see Bingham et al. (1987, p. 43). The mean of the limit law in Theorem 6.2 (resp. Theorem 6.6) is

$$\mu(\varepsilon) = \varepsilon + \int_\varepsilon^\infty (\mathcal{L}(x)/\mathcal{L}(\varepsilon))\,dx < \infty,$$

where $\varepsilon > 0$ (resp. $\varepsilon = 0$). Hence Assertion (b) follows from (a) and Lemma 6.1(d). Since $F(x) = 1 - \mathcal{L}(x)/\mathcal{L}(0)$, its Laplace–Stieltjes transform is $1 - h(s)/h(\infty)$, the right hand side of (6.14). It follows from Lemma 6.1(c) that $\mu(0) = 1/L(\infty)h(\infty)$. Thus (6.15) follows from (6.1) and (6.3). Finally, (6.4) entails $\Psi(1,t) = g(e^{Dt})$ and the last assertion follows from Theorem 6.6 and Lemma 6.2(c). □

We now indicate the effects of starting the CB-process with an arbitrary initial law $\mathcal{M}$ which, without loss of generality, satisfies $\mathcal{M}(\{0\}) = 0$. Let $P_{\mathcal{M}}(\cdot) = \int_0^\infty P_z(\cdot)\mathcal{M}(dz)$, and similarly for $E_{\mathcal{M}}(\cdot)$, and let

$$\Psi_{\mathcal{M}}(t) = 1 - E_{\mathcal{M}}(e^{-Z_t}) = 1 - \widehat{\mathcal{M}}(\Psi(1,t)).$$

In place of the quotient in (6.5), we consider the quotient

$$Q_{\mathcal{M}}(s,t) = \frac{1 - E_{\mathcal{M}}(e^{-sZ_t})}{\Psi_{\mathcal{M}}(t)} = \frac{1 - \widehat{\mathcal{M}}(\Psi(s,t))}{1 - \widehat{\mathcal{M}}(\Psi(1,t))}.$$

Observe that $Q_{\mathcal{M}}(s,t)$ is a cf and, from (6.5) and the definition of regular variation, that it has a limit as $t \to \infty$ iff if $1 - \widehat{\mathcal{M}}(s)$ is regularly varying, i.e., $1 - \widehat{\mathcal{M}}(s) = s^{\delta}\overline{L}(1/s)$ where $0 \le \delta \le 1$ and $\overline{L}$ is slowly varying at infinity. If $\mathcal{M}$ has a finite first moment then $\delta = 1$. We assume that $\widehat{\mathcal{M}}$ is regularly varying, and that $\delta > 0$ to avoid degenerate outcomes.

It follows from (6.5) and the continuity theorem for cf's that $Q_{\mathcal{M}}(s,t) \to (h(s))^{\delta}$ and that the limit is a cf,

$$(h(s))^{\delta} = \int_0^\infty (1 - e^{-sx})\lambda(dx;\delta),$$

where $\lambda(dx;\delta)$ is a Lévy measure. This outcome yields generalizations of the above limit theorems. For just one example, the second assertion of Theorem 6.2 generalizes to

$$\lim_{t\to\infty} P_{\mathcal{M}}(Z_t \le x | Z_t > \varepsilon) = 1 - \mathcal{L}(x;\delta)/\mathcal{L}(\varepsilon;\delta),$$

where $\mathcal{L}(x;\delta) = \lambda((x,\infty);\delta)$.

A special röle is played by initial laws $\mathcal{M}$ which are quasi-stationary, i.e.

$$P_{\mathcal{M}}(Z_t \le x | Z_t > 0) \equiv \mathcal{M}([0,x]), \qquad (t \ge 0).$$

Here we place no initial restriction on $\mathcal{M}$ but, letting $x \to 0+$, we see that $\mathcal{M}(\{0\}) = 0$. Next, if (2.8) does not hold, then the definition is tantamount to assuming the existence of a finite stationary measure, and Lemma 5.1 shows that is not possible. Hence we assume that (2.8) holds, and then the definition can be expressed as

$$P_{\mathcal{M}}(Z_t \in A | T > t) = \mathcal{M}(A), \tag{6.16}$$

for all $t \ge 0$ and Borel sets $A \in (0,\infty)$, and $\mathcal{M}(\{0\}) = 0$. The following result gives necessary conditions for quasi-stationarity. Let $H_{\mathcal{M}}(t) = P_{\mathcal{M}}(T > t) = 1 - \widehat{\mathcal{M}}(\Psi(t))$.

Lemma 6.4. *Suppose* (2.8) *holds and that* $\mathcal{M}$ *is a quasi-stationary law. Then there is a number* $\gamma \in (0, -D]$ *such that (i)* $H_{\mathcal{M}}(t) = e^{-\gamma t}$*; and (ii),* $\mathcal{M}$ *is* γ*-invariant, i.e., for all Borel sets* A*,*

$$P_{\mathcal{M}}(Z_t \in A) = e^{-\gamma t}\mathcal{M}(A), \qquad (t \ge 0). \tag{6.17}$$

Proof. It follows from (2.3) that the Laplace-Stieltjes transform version of (6.16) is

$$\widehat{\mathcal{M}}(\Psi(s,t)) - \widehat{\mathcal{M}}(\Psi(t)) = \widehat{\mathcal{M}}(s)H_{\mathcal{M}}(t).$$

If $u > 0$, then letting $s = \Psi(u)$ in this identity yields the Cauchy functional equation

$$H_{\mathcal{M}}(t+u) = H_{\mathcal{M}}(t)H_{\mathcal{M}}(u).$$

Since $H_{\mathcal{M}}(t)$ is monotonic, this equation has the asserted exponential form for some $\gamma \in [0,\infty]$. The extreme values are not possible since, otherwise, $\mathcal{M}$ is degenerate at zero or infinity. The invariance relation (6.17) follows by multiplying (6.16) through by $H_{\mathcal{M}}(t)$.

Finally, observe for each $x > 0$ that $P_z(Z_t > x)$ is non-decreasing in z, a stochastic monotonicity property shared by all Markovian branching processes. Choose numbers $0 < a < b < \infty$ such that $\mathcal{M}((a,b]) > 0$. It follows from (6.17) that

$$e^{-\gamma t}\mathcal{M}((x,\infty)) \geq \int_a^b P_z(Z_t > x)\mathcal{M}(dz) \geq P_a(Z_t > x)\mathcal{M}((a,b]).$$

However, Theorem 6.3, Lemma 6.2(c) and (6.4) imply that that $P_a(Z_t > x)$ is asymptotically proportional to $e^{Dt}/L^*(e^{-Dt})$. Hence $\gamma \leq -D$. □

The next result asserts the existence of a unique quasi-stationary law for each admissible value of γ, and that each such law arises as a limiting conditional law obtained from an initial law $\mathcal{M}$ whose Laplace-Stieltjes transform is regularly varying at the origin, as above. We state this condition directly in terms of $\mathcal{M}$. The proof is very similar to that used by Asmussen and Hering (1983) for the MBP, and hence we omit it.

Theorem 6.8. *Suppose $\mathcal{M}$ is an initial law satisfying $\mathcal{M}(0) = 0$, and that (2.8) holds. Then the conditioned law $P_{\mathcal{M}}(Z_t \leq x | T > t)$ has a non-defective limit law iff, for some $0 < \delta \leq 1$, $x^{-\delta}\mathcal{M}((x,\infty))$ is SV. If this condition holds then the limit DF F_δ has the Laplace–Stieltjes transform*

$$\begin{aligned} \widehat{F_\delta}(s) &= 1 - (h(s)/h(\infty))^{\delta} \\ &= 1 - \exp\left(-\delta D \int_s^{\infty} du/\psi(s)\right). \end{aligned} \quad (6.18)$$

Its mean is infinite if $\delta < 1$. The measure on $(0,\infty)$ induced by F_δ is $(-\delta D)$-invariant for the transition semigroup of the CB-*process. Finally, the family $(F_\delta : 0 < \delta \leq 1)$ comprises all the quasi-stationary laws.*

Clearly the above index parameter $\gamma = -\delta D$. The expression (6.18) for the Laplace–Stieltjes transform of F_δ comes from (6.1). Further properties of this law are given by Pakes (1997b), together with an elementary demonstration, independent of the branching process context, that $\widehat{F_\delta}$ actually *is* the Laplace–Stieltjes transform of a DF.

We end this section with two examples.

Example 6.1. Let $\nu = 0$ in (2.1) and suppose $Q(\mathbb{R}_+) < \infty$, i.e., the jump component of the underlying Lévy process is compound-Poisson. As shown by by Pakes and Trajstman (1985), sample paths comprise exponentially decaying segments connected by random positive jumps, and there are finitely many of these in $\mathbb{R}_+$ if $D = -\kappa + \rho\eta \leq 0$. If the compound-Poisson component has rate ρ and an Exp(η^{-1}) jump law, then $D < 0$ iff $\alpha = \rho\eta/\kappa < 1$. Some calculation yields

$$h(s) = s\left(\frac{1-\alpha+\eta}{1-\alpha+\eta s}\right)^{\alpha}$$

and

$$\mathcal{L}(x) = (2-\rho/\kappa)^{\alpha} x^{\alpha-1} e^{-(1-\rho/\kappa)x}/\Gamma(\alpha).$$

Example 6.2. Set $\nu = 0$ and take $Q(\cdot)$ as in Example 5.1. We also take $a > 0$ to ensure $D < 0$, and we rewrite the cf as

$$\psi(s) = -as - (b/\delta)s^{1+\delta}$$

where $0 < \delta \leq 1$. These choices imply (2.8). Then, with $\zeta = b/a\delta$,

$$\Psi(s,t) = \left[(\zeta + s^{-\delta})e^{a\delta t} - \zeta\right]^{-1/\delta},$$

(see Bingham (1976, p. 229)) and by direct evaluation of the limit (6.5) we find that

$$h(s)/s = \left(\frac{1+\zeta}{1+\zeta s^{\delta}}\right)^{1/\delta}.$$

Hence $L(\infty) = (1+\zeta)^{-1/\delta}$, and the limit law in Theorem 6.6 is a scaled Linnik law, i.e., the law of $\zeta^{1/\delta}\Gamma_{1/\delta}S_{\delta}$. If $\delta = 1$ this reduces to $(b/a)\mathcal{E}$.

If the initial distribution $\mathcal{M}$ in Theorem 6.8 has a finite first moment, then the conditional limit law there has the Laplace–Stieltjes transform

$$\widehat{F}(s) = 1 - \left(\frac{\zeta s^{\delta}}{1+\zeta s^{\delta}}\right)^{1/\delta}.$$

Setting $\delta = 1$ and replacing b with $\nu/2$ gives the Exp($2a/\nu$) law as the limiting conditional law for the subcritical CB–diffusion process.

7 The Last Large Jump

Assume in this section that Q charges $(0,\infty)$ and, until further notice, that $-\infty < D \leq \infty$. We begin by recalling Lamperti's (1967) construction of the CB-process as a random time transformation of a Lévy process having the cf (2.1). Let $\mathcal{G}_t = \sigma\{L_u : u \leq t\}$, and $C = \inf\{t > 0 : L_t = 0\}$ denote the first zero-hitting or crossing time of the Lévy process. Define the integral functional $T(t) = \int_0^t du/L_u$ if $t < C$, $= \infty$ if $t \geq C$. It is continuous for all t and strictly increasing for $t < C$, and $T(C-) < \infty$ iff (2.8)) holds. The inverse process, $\tau(t) = \inf\{u : T(u) > t\}$ is defined for $t < \infty$, and $\tau(t)$ is a stopping time for $(\mathcal{G}_t)$. The CB-process is given as the random time transformation $Z_t = L_{\tau(t)}$.

Fix $\varepsilon > 0$ and let $q(\varepsilon) = Q((\varepsilon,\infty))$ which we assume is positive. Let $(\Delta L)_t = L_t - L_{t-}$ be the jump size of (L_t) at time t, and let

$$J_t(\varepsilon) = \sum_{u \le t} (\Delta L)_u 1((\Delta L)_u > \varepsilon),$$

which is the sum over $[0,t]$ of (upward) jumps whose size exceeds ε. With probability 1, this sum has only finitely many terms, and $(J_t(\varepsilon))$ is a compound Poisson process with rate $q(\varepsilon)$ and jumps which have the DF $Q((\varepsilon,x])/q(\varepsilon)$. Next, define the process $(L_t(\varepsilon))$ *via* the decomposition

$$L_t = L_t(\varepsilon) + J_t(\varepsilon). \tag{7.1}$$

So $(L_t(\varepsilon))$ is the process obtained from (L_t) by deleting jumps of size $> \varepsilon$. It is a Lévy process, and it is independent of $(L_t(\varepsilon))$. See Bertoin (1996, §I.1) for background facts. Finally, let $\Upsilon_\varepsilon = \inf\{t : (\Delta Z)_t > \varepsilon\}$ be the time of the first "ε-jump" of the CB-process. We begin deriving the law of Υ_ε by finding the local rate of occurence of ε-jumps for the CB-process. This extends the principal content of Theorem 2.1 in Pakes and Trajstman (1985). Let $\mathcal{F}_t = \sigma\{Z_u : u \le t\}$.

Lemma 7.1. *As $h \to 0+$,*

$$P\big(t < \Upsilon_\varepsilon < t+h \,|\, \mathcal{F}_t \cap \{t < T\}\big) = q(\varepsilon) Z_t h + o(h). \tag{7.2}$$

Proof. Letting $K_\varepsilon = \inf\{t : \Delta L_t > \varepsilon\}$, it follows from the random time change construction and the Markov property of (L_t) that the left-hand side of (7.2) equals

$$P\Big(\tau(t) < K_\varepsilon < \tau(t+h) \,|\, \mathcal{G}_{\tau(t)} \cap \{\tau(t) < C\}\Big) = q(\varepsilon)\big(\tau(t+h) - \tau(t)\big)(1 + o(1)).$$

For any $0 < t < T$ the definition of τ yields

$$h = \int_{\tau(t)}^{\tau(t+h)} du/L_u = \frac{\tau(t+h) - \tau(t)}{L_{\tau(t)}}(1 + o(1)), \tag{7.3}$$

where the remainder term is a consequence of the pathwise right-hand continuity of (L_t). Consequently $\tau(t+h) - \tau(t) = Z_t h(1 + o(1))$ when $t < T$. □

We shall now find an expression for the survivor function $N(t|z) := P_z(\Upsilon_\varepsilon > t)$. It is continuous for $t \ge 0$. Let $\psi_\varepsilon(s)$ denote the cf of $(L_t(\varepsilon))$, and let $(Z_t(\varepsilon))$ be the corresponding CB-process obtained from the random time change construction. Its cf is denoted by $\Psi_\varepsilon(s,t)$. We will use the following general formula for the joint Laplace–Stieltjes transform of Z_t and the path integral $\int_0^t Z_u du$:

$$E_z\left(e^{-sZ_t - \theta \int_0^t Z_u du}\right) = e^{-z\Psi(s,\theta,t)}$$

where

$$\frac{\partial \Psi}{\partial t} = \theta + \psi(\Psi) \quad \& \quad \Psi(s,\theta,0) = s. \tag{7.4}$$

We write $\Psi_\varepsilon(s,\theta,t)$ for the joint cf obtained from these expressions as applied to the process $(Z_t(\varepsilon))$, $R_\varepsilon(t) := \Psi_\varepsilon(0,q(\varepsilon),t)$, and $r_\varepsilon := R_\varepsilon(\infty)$.

Theorem 7.1. *The survivor function of* Υ_ε *is*

$$N(t|z) = E_z\left[\exp\left(-q(\varepsilon)\int_0^t Z_u(\varepsilon)du\right)\right] = \exp(-zR_\varepsilon(t)),$$

and

$$P_z(\Upsilon_\varepsilon = \infty) = e^{-zr_\varepsilon}, \tag{7.5}$$

and the right-hand side is zero iff $\nu = 0$ *and* $a - \int_0^1 xQ(dx) \geq 0$.

Proof. Observe first that the branching property implies that $N(t|z) = \exp(-zn(t))$ for some function $n(t)$. This follows from a consideration of the independent growth of partitioned elements of the original population mass; the first ε-jump time of the whole coincides with the first such jump among the components of the partition. Note that $n(0) = 0$ and that n is a.e. differentiable in $(0,\infty)$.

Next, if $h < t$ and $\Upsilon_\varepsilon > t$, then there is no ε-jump in $(0,h)$, an event which we denote by $A(h,\varepsilon)$. Hence

$$N(t|z) = E_z\left[N(t-h|Z_h);A(h,\varepsilon)\right] = (1-q(\varepsilon)h+o(h))E_z\left[N(t-h|Z_h)|A(h,\varepsilon)\right].$$

If $A(h,\varepsilon)$ occurs, then $Z_h = Z_h(\varepsilon)$, so on this event we have for t a.e. that

$$N(t-h|Z_h) = N(t|Z_h(\varepsilon)) - h\frac{\partial}{\partial t}N(t|Z_h(\varepsilon)) + o(h) = N(t|Z_h(\varepsilon)) + hn'(t)N(t|z) + o(h),$$

and hence

$$\frac{1}{h}E_z\left[N(t|z) - N(t|Z_h(\varepsilon))\right] = zN(t|z)(-q(\varepsilon)+n'(t)) + o(1). \tag{7.6}$$

The left-hand side equals $(1/h)[\exp(-zn(t)) - \exp(-z\Psi_\varepsilon(n(t),h))]$. So using the backward equation for $\psi_\varepsilon(s)$, we see that the limit as $h \to 0+$ of the left-hand side of (7.6) is $z\psi_\varepsilon(n(t))N(t|z)$, whence

$$n'(t) = q(\varepsilon) + \psi_\varepsilon(n(t)) \;\;\&\;\; n(0) = 0. \tag{7.7}$$

Comparison with (7.4) shows that $n(t) = \Psi(0,q(\varepsilon),t)$ for all $t \geq 0$, giving the first assertion.

Since the stochastic integral $\int_0^t Z_u(\varepsilon)du$ is non-decreasing in t it follows that $\Psi(0,\theta,t)$ increases with t. Recollecting that ψ_ε eventually is negative and decreasing, it then follows from (7.4) that $n(t) \to r_\varepsilon > 0$, the unique positive solution of $q(\varepsilon) + \psi(s) = 0$. Thus (7.5) follows upon letting $t \to \infty$ in the first assertion. Graphical considerations and the concavity of ψ imply that $r_\varepsilon < \infty$ if $\psi(\infty) = -\infty$, and $r_\varepsilon = \infty$ if $\psi(\infty) = \infty$. The second condition occurs iff $P_z(Z_t \to \infty) \equiv 1$. Thus the final assertion follows from the remarks ending §2. □

Our next step is to determine the law of the epoch $\mathcal{L}_\varepsilon$ of the *last* ε-jump. Recall that ρ is the largest real zero of ψ, and that $P_z(Z_t \to \infty) = 1 - e^{-z\rho}$. We determine $P_z(\mathcal{L}_\varepsilon = \infty)$ as follows. The decomposition (7.1) implies that almost all sample paths of (L_t) contain infinitely many ε-jumps. The CB-process paths in the event $\{Z_t \to \infty\}$ are in one-one correspondence with the paths of (L_t) which never hit $\{0\}$. Consequently $P_z(\mathcal{L}_\varepsilon = \infty | Z_t \to \infty) = 1$.

Recall that $(\overline{Z}_t)$ denotes the non-supercritical CB-process which is equal in law to (Z_t) conditioned on the event $\{Z_t \to 0\}$. Almost all sample paths of $(\overline{Z}_t)$ have only finitely many ε-jumps, for if not, then the corresponding (first) positive excursion of the generating Lévy process would explode to ∞ before it crosses the origin. Consequently, $\overline{\mathcal{L}}_\varepsilon = \infty$ iff the conditioned CB-process has no ε-jumps. Using (7.5), it follows that

$$P_z(\mathcal{L}_\varepsilon = \infty | Z_t \to 0) = P_z(\overline{\Upsilon}_\varepsilon = \infty) = e^{-z\overline{r}_\varepsilon},$$

where $\overline{r}_\varepsilon$ is the unique positive solution of $\overline{\psi}(s) + \overline{q}(\varepsilon) = 0$, and $\overline{q}(\varepsilon) = \int_\varepsilon^\infty e^{-\rho x} Q(dx)$. The law of total probability now implies that

$$P_z(\mathcal{L}_\varepsilon = \infty) = 1 - e^{-z\rho} + e^{-z(\overline{r}_\varepsilon + \rho)}.$$

It is easily seen from a sketch graph that if $D > 0$ then $\overline{r}_\varepsilon + \rho < r_\varepsilon$.

The above argument shows that $\mathcal{L}_\varepsilon < \infty$ only if $Z_t \to 0$, and hence we will assume from now on that $D \leq 0$, in which case $\rho = 0$, so

$$P_z(\mathcal{L}_\varepsilon = \infty) = e^{-zr_\varepsilon}. \tag{7.8}$$

Theorem 7.2. *If $D \leq 0$, then the law of $\mathcal{L}_\varepsilon$ is given by*

$$\lambda(t|z) := P_z(\mathcal{L}_\varepsilon \leq t) = e^{-z\Psi(r_\varepsilon, t)} - e^{-zr_\varepsilon}.$$

In particular

$$P_z(t < \mathcal{L}_\varepsilon < \infty) = 1 - e^{-z\Psi(r_\varepsilon, t)}. \tag{7.9}$$

Proof. We begin by showing that there is a continuous monotone function $V(t) > 0$ such that

$$\lambda(t|z) = e^{-zV(t)} - e^{-zr_\varepsilon}. \tag{7.10}$$

To see this, partition the parent population of mass z into n elements of mass z/n, $n = 1, 2, \ldots$. For each n, let $\mathcal{L}_\varepsilon(1,n), \ldots, \mathcal{L}_\varepsilon(n,n)$ be the times of last ε-jump for the independent CB-processes initiated by the partition elements. Then there *is* a last ε-jump iff for some $k = 1, \ldots, n$, k of the $\mathcal{L}_\varepsilon(j,n)$ are $\leq t$ and the other $n-k$ of them equal ∞. It follows from (7.8) that

$$\lambda(t|z) = \sum_{k=1}^{n} \binom{n}{k} (\lambda(t|z/n))^k e^{-(zr_\varepsilon/n)(n-k)} = \left[\lambda(t|z/n) + e^{-zr_\varepsilon/n}\right]^n - e^{-z\rho}.$$

Rewriting the right-hand side as $e^{-zr_\varepsilon}\left[(1 + \lambda(t|z/n)e^{zr_\varepsilon/n})^n - 1\right]$, we see that $\lambda(t|z/n)$ is asymptotically proportional to z/n. Denoting the constant of proportionality by $r_\varepsilon - V(t)$ we thus obtain

$$\lambda(t|z) = \left[1 - zr_\varepsilon/n + (z/n)(r_\varepsilon - V(t)) + o(n^{-1})\right]^n - e^{-zr_\varepsilon},$$

and (7.10) now follows by letting $n \to \infty$.

A straightforward backward decomposition suffices to determine the form of $V(t)$, as follows. Let $\ell(t|z) = P_z(t < \mathcal{L}_\varepsilon < \infty)$. Then (7.8) and (7.10) yield $\ell(t|z) = 1 - e^{-zV(t)}$. But, for t a.e.,

$$\begin{aligned}\ell(t|z) &= E_z\left[\ell(t-h, Z_h)\right] = E_z\left[\ell(t|Z_h) - Z_h hV'(t)e^{-Z_h V(t)}\right] + o(h)\\ &= E_z\left[\ell(t|Z_h) - zhV'(t)e^{-zV(t)}\right] + o(h).\end{aligned}$$

However,

$$\begin{aligned}\lim_{h\to 0+}\left(\ell(t|z) - E_z(\ell(t|Z_h))\right)/h &= \lim_{h\to 0+}\left(e^{-z\Psi(V(t),h)} - e^{-zV(t)}\right)/h\\ &= -z\psi(V(t))e^{-zV(t)},\end{aligned}$$

and it follows that $V'(t) = \psi(V(t))$ for $t > 0$, which is the backward differential equation for (Z_t). The assertions follow since $V(0) = r_\varepsilon$. □

If $q_2 < \infty$ we can let $\varepsilon = 0$, and then $\mathcal{L}_0$ is the time of last positive jump, considered by Pakes (1988) under the further assumption $\nu = 0$. In this case $\psi_0(s) = -\kappa s$ and hence $r_\varepsilon = q_2/\kappa$, agreeing with Pakes (1988) (after noting that the event $\{T > t\}$ in this reference should read $\{t < T < \infty\}$).

The whole development of §§4,6 can be duplicated by conditioning Z_t on events implying that $\mathcal{L}_\varepsilon > t$. Thus, using (7.9),

$$\begin{aligned}E_z\left[e^{-sZ_t}; t+\tau < \mathcal{L}_\varepsilon < \infty\right] &= E_z\left[e^{-sZ_t}P_{Z_t}(\tau < \mathcal{L}_\varepsilon < \infty)\right]\\ &= E_z\left[e^{-sZ_t}\left(1 - e^{-Z_t\Psi(r_\varepsilon,\tau)}\right)\right]\\ &= \left[e^{-z\Psi(s,t)} - e^{-z\Psi(s+\Psi(r_\varepsilon,\tau),t)}\right].\end{aligned} \tag{7.11}$$

This can be used to prove that if $D \le 0$, if $t \to \infty$, and if τ varies with t so that it remains bounded away from zero, then

$$E_z\left[e^{-sZ_t}|t+\tau < \mathcal{L}_\varepsilon < \infty\right] \sim \frac{\Psi(s+\Psi(r_\varepsilon,\tau),t) - \Psi(s,t)}{\Psi(\Psi(r_\varepsilon,\tau),t)}. \tag{7.12}$$

This estimate can be used in a similar way to the representation in Lemma 4.1 to obtain limit theorems parallel to many in §4. We mention only a few instances.

Suppose $D = 0$ and (4.2) holds. If $0 \le \alpha \le \infty$ and $\tau/t \to \alpha$ we infer that

$$b_t\Psi(r_\varepsilon,\tau) = (b_t/b_\tau)b_\tau\Psi(r_\varepsilon,\tau) \to c = \alpha^{-1/\delta}\Omega(\mathbb{R}_+).$$

Similar working which gave the main results in §4 leads to

$$\lim_{t\to\infty} P_z\left(Z_t \le xb_t | t+\tau < \mathcal{L}_\varepsilon < \infty\right) = G(x,\alpha),$$

where $\widehat{G}(s,\alpha)$ is given by (4.13) and (4.4). Thus we see in the critical case that the limit theorems obtained by conditioning on $\mathcal{L}_\varepsilon$ coincide with those for the case where (2.8) is valid and we condition on the extinction time T.

If $D < 0$ and τ is fixed, then (6.5) applied to (7.12) yields the limiting Laplace–Stieltjes transform $[h(s+\Psi(r_\varepsilon,\tau)) - h(s)]/h(\Psi(r_\varepsilon,\tau))$. Using (6.6) and (7.9) shows that

$$\lim_{t\to\infty} P_z\big(Z_t \le x | t+\tau < \mathcal{L}_\varepsilon < \infty\big) = \frac{\int_0^x P_y(\tau < \mathcal{L}_\varepsilon < \infty)\lambda(dy)}{\int_0^\infty P_y(\tau < \mathcal{L}_\varepsilon < \infty)\lambda(dy)}.$$

The denominator can be expressed as $h(r_\varepsilon)P(\mathcal{E}_D > \tau)$. Thus an analogue of Theorem 6.7 is immediately evident.

Next, if $D \le 0$ and $\tau \to \infty$ in (7.12), then as in Theorem 4.7,

$$\lim_{t\to\infty} E_z\left[e^{-sZ_t} | t+\tau < \mathcal{L}_\varepsilon < \infty\right] = e^{-Dt}\psi(\Psi(s,t))/\psi(s),$$

and analogues of Theorems 4.8 and 6.5 follow.

Finally we consider $Z_{\mathcal{L}_\varepsilon}$ under the condition that $\mathcal{L}_\varepsilon$ is large and finite. The following result generalizes the principal calculation used for proving Theorem 5.1 in Pakes (1988).

Theorem 7.3. *If* $D \le 0$ *then*

$$E_z\left[e^{-sZ_{\mathcal{L}_\varepsilon}}; t < \mathcal{L}_\varepsilon < \infty\right] = \frac{\widehat{Q}_\varepsilon(r_\varepsilon+s)}{-\psi(r_\varepsilon+s)}\left[1 - e^{-z\Psi(r_\varepsilon+s,t)}\right]. \tag{7.13}$$

Proof. The event $\{\mathcal{L}_\varepsilon \in (t,t+dt)\}$ occurs iff there is an ε-jump during $(t,t+dt)$ and none occur subsequently. The size of the process immediately after the last ε-jump is $Z_{\mathcal{L}_\varepsilon} = Z_t + J$ where J is the size of a single ε-jump and it is independent of $\mathcal{F}_t$. Ignoring events whose probability is of order $o_p(dt)$, it follows from Lemma 7.1 that the the conditional probability of an ε-jump in $(t,t+dt)$, given $\mathcal{F}_t$, is $q(\varepsilon)Z_t dt$ and, from (7.5), the conditional probability of no ε-jump after $t+dt$ is $e^{-r_\varepsilon(Z_t+J)}$. Putting these together yields

$$\begin{aligned} E_z\left[e^{-sZ_{\mathcal{L}_\varepsilon}}; \mathcal{L}_\varepsilon \in (t,t+dt)\right] &= q(\varepsilon)dt E_z\left[Z_t e^{-(r_\varepsilon+s)(Z_t+J)}\right] \\ &= dt\widehat{Q}_\varepsilon(r_\varepsilon+s)\left(-\frac{\partial}{\partial s}\right)E_z\left[e^{-(r_\varepsilon+s)Z_t}\right] \\ &= dt\widehat{Q}_\varepsilon(r_\varepsilon+s)z\Psi'(r_\varepsilon+s,t)e^{-z\Psi(r_\varepsilon+s,t)} \\ &= dt\,\frac{\widehat{Q}_\varepsilon(r_\varepsilon+s)}{-\psi(r_\varepsilon+s)}\left(-\frac{\partial}{\partial t}\right)e^{-z\Psi(r_\varepsilon+s,t)}, \end{aligned}$$

where the last line follows from (4.15) and the backward differential equation. Since $\Psi(s,\infty) = 0$, the assertion follows by integrating. □

Note that setting $s = 0$ in (7.12) yields (7.9) because

$$\psi(r_\varepsilon) = \psi_\varepsilon(r_\varepsilon) + q(\varepsilon) - \widehat{Q}_\varepsilon(r_\varepsilon) = -\widehat{Q}_\varepsilon(r_\varepsilon).$$

Hence the Laplace–Stieltjes transform of the conditional law of $Z_{\mathcal{L}_\varepsilon}$, given that $t < \mathcal{L}_\varepsilon < \infty$, is obtained by dividing (7.9) into (7.13). The following limit theorem follows easily from Lemma 5.2 if $D = 0$, and (6.5) if $D < 0$.

Theorem 7.4. *If* $D \leq 0$ *then* $\lim_{t\to\infty} P_z(Z_{\mathcal{L}_\varepsilon} \leq x | t < \mathcal{L}_\varepsilon < \infty) = A(x)$ *exists as a non-defective* DF *and*

$$\widehat{A}(s) = \frac{\widehat{Q}_\varepsilon(r_\varepsilon + s)}{-\psi(r_\varepsilon + s)} \times \begin{cases} 1 & \text{if } D = 0, \\ \dfrac{h(r_\varepsilon + s)}{h(r_\varepsilon)} & \text{if } D < 0. \end{cases}$$

This result has the same algebraic form as the limit theorem assertion in Theorem 5.1 of Pakes (1988), proved there under the further assumptions $\nu = 0$, $q_2 < \infty$ and $\varepsilon = 0$. If $D = \nu = 0$ and $Q(dx) = Ae^{-\mu x}dx$, with A a positive constant, then routine calculation shows that the limit law can be expressed in random variable terms as $\varepsilon + (\mathcal{E} + \mathcal{E}')/\rho$ where $\mathcal{E}$ and $\mathcal{E}'$ are independent and have a standard exponential law. Even for this simple case, however, r_ε is determined as the unique positive solution of the nonlinear equation $s^2 = \mu^2 e^{-(\mu+s)\varepsilon}$.

References

[1] Asmussen, S.; Hering, H. *Branching Processes*. Birkhäuser: Boston, MA, 1983.

[2] Bertoin, J. *Lévy Processes*. Cambridge University Press: Cambridge, 1996.

[3] Bertoin, J. Cauchy's principal value of local times of Lévy processes with no negative jumps via continuous branching processes. *Electronic J. Prob.* 2, 6, 1–12.

[4] Bingham, N. H. Continuous branching processes and spectral positivity. *Stoch. Processes Appl.* 1976, *4*, 217–242.

[5] Bingham, N.H.; Goldie. C.M.; Teugels, J.L. *Regular Variation.* Cambridge University Press: Cambridge, 1987.

[6] Bühler, W. J. Some results on the behaviour of branching processes. *Theor. Prob. Appl.* 1968, *13*, 52–64.

[7] Conner, H. E. A note on limit theorems for Markov branching processes. *Proc. Amer. Math. Soc.* 1967, *18*, 76–86.

[8] Duquesne, T.; Le Gall, J.-F. Random Trees, Lévy Processes and Spatial Branching Processes. Astérisque 281, 2002.

[9] Ginzberg, L. R.; Slobodkin, L. B.; Johnson, K.; Bindman, A. G. Quasi-extinction probabilities as a measure of impact on population growth. *Risk Anal.* 1982, *2*, 171–181.

[10] Grey, D.R. Asymptotic behaviour of continuous time, continuous state-space branching processes. *J. Appl. Prob.* 1974, *11*, 669–677.

[11] Grimvall, A. On the convergence of sequences of branching processes. *Ann. Prob.* 1974, *2*, 1027–1045.

[12] Harn, K. van; Steutel, F.W. Stability equations for processes with stationary independent increments using branching processes and Poisson mixtures. *Stoch. Processes. Appl.* 1993, *45*, 209–230.

[13] Jiřina, M. Stochastic branching processes with continuous state space. *Czech. Math. J.* 1958, *8*, 292–312.

[14] Kallenberg, P. J. M. *Branching Processes with Continuous State space.* Mathematisch Centrum: Amsterdam, 1979.

[15] Kim. Y. Mechanisms of symptom in Bartoszyński's virus model. *Math. Biosci.* 1998, *153*, 63–78.

[16] Lambert, A. Coalescence times for the branching process. *Adv. Appl. Prob.* 2003, *35*, 1071–1089.

[17] Lambert, A. . Quasi-stationary distributions and the continuous-state branching process conditioned to be never extinct. *Prob. Theory Rel. Fields*. 2008, to appear.

[18] Lamperti, J. Continuous state branching processes. *Bull. Amer. Math. Soc.* 1967a, *73*, 382–386.

[19] Lamperti, J. The limit of a sequence of branching processes. *Z. Wahrscheinlichkeitsth. verw. Geb.* 1967b, *7*, 271–288.

[20] Li, Z.-H. Asymptotic behaviour of continuous time and state branching processes. *J. Austral. Math. Soc. (Series A)*, 2000, *68*, 68–84.

[21] Morris, W. F.; Doak, D. F. *Quantitative Conservation Biology.* Sinauer Associates, Inc., Sunderland, MA, 2002.

[22] Ogura, Y. Spectral representation for branching processes on the real half line. *Publ. RIMS Kyoto Univ.* 1970, *5*, 423–441.

[23] Pakes, A. G. Some limit theorems for Jiřina process. *Periodica Math. Hung.* 1979, *10*, 55-66.

[24] Pakes, A. G. Some limit theorems for continuous-state branching processes. *J. Austral. Math. Soc. (Series A)* 1988, *44*, 71–87.

[25] Pakes, A. G. Characterization by invariance under length-biasing and random scaling. *J. Statist. Planning Inf.* 1997a, *63*, 285–310.

[26] Pakes, A. G. *On the recognition and structure of probability generating functions* In Classical and Modern Branching Processes, Athreya, K. B.; Jagers, P. Editors, Springer-Verlag: New York, 1997b, pp. 263–284.

[27] Pakes, A. G. Mixture representations for symmetric generalized Linnik laws. *Statist. Prob. Letters.* 1998a, *37*, 213–221.

[28] Pakes, A. G. Revisiting conditional limit theorems for the mortal simple branching process. *Bernoulli*, 1999, *5*, 969–998.

[29] Pakes, A. G.; Trajstman, A. C. Some properties of continuous-state branching processes, with applications to Bartoszyński's virus model. *Adv. Appl. Prob.* 1985, *17*, 23–41.

[30] Pratt, J. W. On interchanging limits and integrals. *Ann. Math. Statist.* 1960, *31*, 74–77.

[31] Seneta, E. Some supplementary notes on one-type continuous-state branching processes. *Z. Wahrscheinlichkeitsth. verw. Geb.* 1976, *34*, 87–98.

[32] Seneta, E.; Vere-Jones, D. On the asymptotic behaviour of subcritical branching processes with continuous state space. *Z. Wahrscheinlichkeitsth. verw. Geb.* 1968, *10*, 212–225.

[33] Seneta, E.; Vere-Jones, D. On a problem of M. Jǐrina concerning continuous state branching processes. *Czech. Math. J.* 1969, *19*, 277–283.

[34] Sevast'yanov, B. A. The theory of branching random processes. *Uspehi Mat. Nauk* 1951, *6*, 47–99.

[35] Shurenkov, V. M. Some limit theorems for branching processes with the continuous space of states. *Teor. Verojatnost. i Mat. Statist. (Kiev)*, 1973, *9*, 167–172 (In Russian).

[36] Slack, R. S. A branching process with mean one and possibly infinite variance. *Z. Wahrscheinlichkeitsth. verw. Geb.* 1968, *9*, 139–145.

[37] Warren, J. *Branching processes, the Ray-Knight theorem, and sticky Brownian motion* In Séminaire de Probabilités XXXI, Azéma, J.; Emery, M.; Yor, M. Editors, Springer-Verlag: Berlin, 1997.

[38] Yaglom, A. M. Certain limit theorems of the theory of branching processes. *Doklady* 1947, *56*, 795–798 (In Russian).

[39] Zhang, Y. C. A limit theorem for the continuous state branching process. *Stoch. Processes. Appl.* 1983, *14*, 175–186. Corrigendum in 15, 327.

[40] Zolotarev, V. M. More exact statements of several theorems in the theory of branching processes. *Theor. Prob. Appl.* 1957, *2*, 245–253.

In: Records and Branching Processes ISBN 978-1-60456-526-3
Editors: M. Ahsanullah, G. P. Yanev, pp. 105-126

Chapter 5

LARGE FAMILIES AND EXCEEDANCES IN BRANCHING PROCESSES

Ibrahim Rahimov*
Department of Mathematical Sciences
King Fahd University of Petroleum and Minerals
P.O. Box 1339, Dhahran, 31261, Saudi Arabia
and
Romanovski Institute of Mathematics
Uzbek Academy of Sciences
700143 Tashkent, Uzbekistan

Abstract

We review some recent results on exceedances and genealogy of branching processes. Define a family in a branching population as a subpopulation of individuals living at time t and having a common ancestor at time $\tau, \tau < t$. We present limit theorems for the number of large families when both τ and t tend to infinity. We also show that if $\tau = 0$ and the underline branching process starts with large number of ancestors, then the large family process is approximated in the Skorohod topology by processes with independent increments. We also present limit theorems for the index of the first process in the sequence of branching processes which exceeds a threshold.

Key words and phrases. Reduced branching processes, exceedance level, sequence of branching processes, limit theorems.

1 Introduction

Although the population size is the main object of investigation in the theory of branching processes, there is a number of other variables related to the genealogy of the processes, which are also of interest. These include: the first exit time of an interval, the distance to the closest common ancestor, the number of pairs having the same number of descendants, reduced processes etc. Here we focus our attention on the large families in a branching

*E-mail address: rahimov@kfupm.edu.sa

population. Our discussion is an attempt of reviewing recent results scattered in different publications.

Let us fix τ to be the time (generation index) when the common ancestor of the family lives. Define a family at time t as a subpopulation of individuals who have a common ancestor at time τ ($\tau < t$). We say that the family is large, if its size is greater than a given threshold $\theta = \theta(t-\tau)$. In this paper we study the asymptotic behavior of the number of large families, when both τ and t tend to infinity.

In Section 5, we present limit theorems for the number of large families in a simple branching process with n types of individuals. In multitype processes a large family can be defined in different manners. For instance, we may consider a family to be large, if the number of members of at least one type is greater than a given threshold function. A more general definition of a large family is based on certain linear combinations of the number of members of different types. More precisely, we say that a family is large, if

$$\sum_{i=1}^{n} a_i X_i^{(\alpha)}(t-\tau) > \theta(t-\tau),$$

where a_i, $i = 1,\dots,n$ are some nonnegative numbers and $X_i^{(\alpha)}(t-\tau)$, $i = 1,\dots,n$ is the number of type i descendants at time t of a common ancestor α living at time τ. Note that this general definition includes the case of families with large total number of members, i.e., $a_i = 1$ for all $i = 1,\dots,n$.

In Section 4 we consider the number of large families defined as follows. Let $\nabla \subseteq \{1,2,\dots,n\}$ and $\theta(t) = (\theta_1(t),\theta_2(t),\dots,\theta_n(t))$ be a n-dimensional vector of nonnegative functions. We say that a family is large, if the number of its members of type $i \in \nabla$ exceeds $\theta = \theta_i(t-\tau)$. Note that different subsets ∇ generate different large families in the sense of types of their members.

It is clear that the number of large families at time t is related to the number of highly productive individuals living at time τ. Therefore, we study the number of highly productive ancestors in a large population. Define $S(n,t)$ to be the number of those among the n ancestors (living at time zero) having more than $\theta(t)$ descendants at time t. A sequence of branching processes may be considered as a population model describing the growth in different regions of a country R. Then one can interpret $S(n,t)$ as the number of regions of R whose population at time t exceeds level $\theta(t)$. Another interpretation of $S(n,t)$ is possible in the framework of vertexes of rooted random trees (see [19]). In fact, each realization of the scheme under consideration can be interpreted as a forest containing n rooted trees. Thus, a realization of $S(n,t)$ is the number of trees in the forest having more than $\theta(t)$ vertexes at level t.

In Section 3 we discuss the convergence of $Y(x,t) = S([m(t)x],t)$ as $t \to \infty$ in the Skorokhod topology, where $x \in [0,\infty)$ and $m(t) \to \infty$ as $t \to \infty$. It turns out that, depending on the rate of divergence of $m(t)$, the process $Y(x,t)$ is approximated by Poisson, Wiener or a process with independent and binomially distributed increments.

We also consider processes with unbounded number of initial ancestors. That is we deal with an infinite sequence of independent Bienaymé–Galton–Watson (BGW) processes $X_i(t)$, $i \geq 1$. Let $\nu(t)$ be the index of the first process in this sequence which population size exceeds $\theta(t)$ at time t. The simple equality

$$\{\nu(t) \leq xt\} = \{S([tx],t) \geq 1\},$$

allows us to apply the results of Section 3 in obtaining limit theorems for $\nu(t)$. These are presented in Section 2, where we study BGW branching processes without and with immigration. We obtain limit results about $\nu(t)$ under different assumptions, such as common or different offspring distributions of the processes and different forms of the threshold $\theta(t)$. We cover all three cases of subcritical, critical, and supercritical processes.

2 Exceeding a Threshold

Let X_{ti}, $i \in N = \{1,2,\ldots\}$, $k \in N_0 = N \cup \{0\}$, be independent and identically distributed random variables, taking values in the set N_0. We define a BGW process $X(t)$, $t \in N$, by the following recurrence $X(0) = 1$ and

$$X(t) = \sum_{i=1}^{X(t-1)} X_{ti}.$$

Here X_{ti} denote the number of direct descendants of the ith individual existing at time t. Note that the probability distribution of the random variables X_{ti} defines the process $X(t)$ completely.

Let us consider a sequence of BGW processes $\{X_i(t), i \in N\}$ and define the process $\nu(t)$, $t \in N_0$ by the relation

$$\nu(t) = \min\{i : X_i(t) > \theta(t)\},$$

where $\theta(t)$ is a given function. The processes $X_i(t)$, $i = 1,2,\ldots$, can be interpreted as sizes of different populations existing in different regions of a country. Then the process $\nu(t)$ is the index of the first region where the population size exceeds level $\theta(t)$. Instead of the population sizes, we can measure some other random characteristics in the sequence of processes $X_i(t)$, $i = 1,2,\ldots$ For instance, we might be interested in the size of a certain product that the individuals produce during their lifetime (cf. [7]). In this setting, $\nu(t)$ will be the index of the first region whose product size at time t reaches level $\theta(t)$.

The properties of exceedances of given levels by sequences of independent and identically distributed random variables have been studied intensively, see for example [9] and [24] and the references therein. Here we will discuss some asymptotic properties of $\nu(t)$. These properties naturally depend on the criticality of the underline branching processes. As a result we obtain three different limit distributions for $\nu(t)$ (and for properly chosen $(\theta(t))$ in the subcritical, critical, and supercritical cases.

2.1 Sequence of processes with common offspring distribution

Let us consider a sequence of BGW processes $\{X_i(t), i \in N\}$ such that

(a) $X_i(t)$, $i \in N$ are independent for any fixed $t \in N_0$ and $X_i(0) = 1$;

(b) $P\{X_i(1) = k | X_i(0) = 1\} = p_k$, $\sum_{k=0}^{\infty} p_k = 1$.

If we think of $X_i(t)$ as the sizes of populations existing in different regions of a country, then the assumptions (a) and (b) mean that the reproduction processes in different regions are independent but have the same offspring laws.

Denote for $i = 1, 2, \ldots$

$$X(t) \overset{\mathrm{d}}{=} X_i(t) \quad Q(t) = P\{X(t) > 0\}, \quad A = EX(t), \quad f(s) = \sum_{k=0}^{\infty} p_k s^k.$$

The notation $\overset{\mathrm{d}}{=}$ stands for equality in distribution. We introduce the following class of stochastic processes for a given function $\varphi_t(y)$ and distribution $\pi(x)$.

Definition. The process $Z(t)$ belongs to the class $\mathcal{A}[\varphi_t(y), \pi(x)]$ if as $t \to \infty$ there exists a random variable ξ such that

$$\varphi_t[Z(t)] \overset{\mathrm{d}}{\to} \xi, \quad \text{and} \quad P\{\xi \le x\} = \pi(x).$$

First we consider the **subcritical case**. It is known that (see [21]) if $A < 1$, then

$$\lim_{t \to \infty} P\{X(t) = j | X(t) > 0\} = p_j^* \quad (j \ge 1)$$

and the probability generating function $f^*(s)$ of $p_j^*, (j \ge 1)$ satisfies the equation:

$$1 - f^*(f(s)) = A(1 - f^*(s)).$$

Theorem 1. *If $A < 1$ and $\theta(t) = \theta \in N = \{1, 2, \ldots\}$, then $\nu(t) \in \mathcal{A}[Q(t)y, E_1(x)]$, where*

$$E_1(x) = \begin{cases} 0, & x < 0 \\ 1 - e^{-q^*(\theta)x}, & x \ge 0 \end{cases}, \quad q^*(\theta) = \sum_{k=\theta+1}^{\infty} p_k^*.$$

If $A \le 1$, then $Q(t) \to 0$ and if, in addition, $EX(1) \ln X(1) < \infty$ then ([21] p. 56).

$$Q(t) \sim KA^t, \quad 0 < K = \prod_{n=0}^{\infty} B(P(X(n) = 0)) < \infty,$$

where $B(s) = (1 - f(s))/(1 - s)$.

Example 1. Consider the geometric offspring distribution $p_k = p(1-p)^k$, $(k \in N_0, 0 < p < 1)$. In this case $f(s) = p(1 - qs)^{-1}$, $q = 1 - p$ and $A = qp^{-1}$. If $p > 1/2$ then the generating function $f^*(s)$ is

$$f^*(s) = \frac{(1 - q/p)s}{1 - qs/p}.$$

Hence p_k^*, $(k \ge 1)$ is the positive geometric distribution with parameter $1 - A$ and

$$q^*(\theta) = \sum_{k=\theta+1}^{\infty} (1 - A)A^{k-1} = A^{\theta}.$$

We also have $B(s) = p(1 - qs)^{-1}$ and $K = 1 - A$. Thus, from Theorem 1 we obtain the following result.

Corollary 1. *If the offspring distribution is geometric with parameter* $p > 1/2$, *and* $\theta \in N$, *then* $\nu(t)(1-A)A^t$ *converges in distribution to a random variable having exponential density function* $A^\theta \exp\{-A^\theta x\}$, *where* $A = (1-p)/p$.

Now we turn to the **critical case**, i.e., $A = 1$. Assume that the offspring generating function $f(s)$ is given by

$$f(s) = s + (1-s)^{1+\alpha} L(1-s), \tag{2.1}$$

where $\alpha \in (0,1)$ and $L(x)$ is a slowly varying at zero function. Note that under this assumption the offspring mean is one and the offspring variance may not be finite. The case of finite variance is included when $\alpha = 1$ and $L(x)$ has a finite limit. If (2.1) holds, then (see [22]),

$$\lim_{t\to\infty} P\{X(t)Q(t) \le x | X(t) > 0\} = G(x) \quad (x \ge 0), \tag{2.2}$$

where $G(x)$ has the Laplace transform $1-\lambda(1+\lambda^\alpha)^{-1/\alpha}$ and the probability of non-extinction satisfies $Q(t) \sim N(t)t^{-1/\alpha}$, where $N(t)$ is slowly varying at infinity and

$$N^\alpha(t)L(N(t)t^{-1/\alpha}) \to \alpha^{-1}.$$

Theorem 2. *If* $f(s)$ *satisfied* (2.1) *and* $\theta(t) = C/Q(t)$ $(0 < C < \infty)$, *then*

$$\nu(t) \in \mathcal{A}[Q(t)y, E_2(x)],$$

where

$$E_2(x) = 1 - e^{-(1-G(C))x} \quad (x \ge 0)$$

and $G(x)$ *is the limit in* (2.2).

Remarks.

(i) *If the offspring mean is one and* $\sigma^2 < \infty$ *then* $Q(t) \sim 2/\sigma^2 t$ *as* $t \to \infty$ *and* $G(x)$ *is exponential with parameter one. Therefore, it follows from Theorem* 2 *that for* $\theta(t) = \sigma^2 Ct/2$ *the process* $2\nu(t)/\sigma^2 t$, *converges in distribution to a random variable with distribution* $1 - \exp\{-e^{-c}x\}$.

(ii) *Let* (2.1) *holds with* $\alpha = 1$ *and* $L(1-s) = -\log(1-s)$. *Then* $N(t) = (\log t)^{-1}$ *and* $Q(t) \sim (t \log t)^{-1}$. *Thus, by Theorem* 2, *if* $\theta(t) = Ct \log t$ *the process* $\nu(t)/t \log t$, *converges to a random variable with the same as in* (i) *exponential distribution.*

Let us consider the **supercritical case**, i.e., $A > 1$. It is known (cf. [1]) that if

$$A > 1 \quad \text{and} \quad EX(1) \ln X(1) < \infty, \tag{2.3}$$

then $X(t)A^{-t}$ converges with probability one to a random variable W, whose Laplace transform $\varphi(\lambda)$ satisfies

$$\varphi(\lambda) = f(\varphi(\lambda/A)). \tag{2.4}$$

The distribution function $H(x)$ of W is absolutely continuous for $x > 0$ and has an atom q at $x = 0$, where q equals the probability of extinction.

Theorem 3. *If* (2.3) *is satisfied and* $\theta(t) = CA^t$ *where* $(0 < C < \infty)$, *then*

$$\nu(t) \in \mathcal{A}\{y, E_3(H(C),x)\},$$

where

$$E_3(H(C),x) = (1 - H(C)) \sum_{k \le x} (H(C))^{k-1}.$$

It can also be seen that, if $\theta(t) \to \infty$ such that $\theta(t)A^{-t} \to \infty$, then $\nu(t) \to \infty$ in probability and, if $\theta(t)A^{-t} \to 0$, the limiting distribution is $E_3(q,x)$ where q is the probability of extinction.

Example 2. Let the offspring distribution be positive geometric, that is $p_k = p(1-p)^{k-1}$, $(k \ge 1)$ and $p_0 = 0$. Then $f(s) = ps(1-qs)^{-1}$, $q = 1-p$, $A = 1/p$ and equation (2.4) becomes

$$\varphi\left(\frac{\lambda}{p}\right) = \frac{p\varphi(\lambda)}{1-(1-p)\varphi(\lambda)}.$$

It is not difficult to see that the Laplace transform $\varphi(\lambda) = p(p+\lambda)^{-1}$ satisfies the above equation. Hence, the limit W is exponential with density function pe^{-px}. Now Theorem 3 implies the following result.

Corollary 2. *If the offspring distribution is positive geometric with parameter* p *and* $\theta(t) = Cp^{-t}$, *then* $\nu(t)$ *has a positive geometric limit such that*

$$\lim_{t\to\infty} P\{\nu(t) = k\} = e^{-pC}(1-e^{-pC})^{k-1} \quad (k \ge 1).$$

We conclude this section with remarks on the asymptotic behavior of the mean and variance of $\nu(t)$. In the subcritical case $(A < 1)$ if $\theta(t) \equiv \theta$, then as $t \to \infty$

$$E\nu(t) \sim \frac{1}{q^*(\theta)}(Q(t))^{-1} \quad \text{and} \quad Var\nu(t) \sim \frac{1-q^*(\theta)}{(q^*(\theta))^2}(Q(t))^{-1}.$$

For the geometric offspring with $p > 1/2$ (see Example 1) $E\nu(t) \sim \frac{p}{2p-1}\left(\frac{p}{1-p}\right)^{t+\theta}$.

If $X(t)$ is critical $(A = 1)$ with (2.1) and $\theta(t) = C/Q(t)$, then as $t \to \infty$

$$E\nu(t) \sim \frac{1}{1-G(C)}(Q(t))^{-1} \quad \text{and} \quad Var\nu(t) \sim \frac{G(C)}{(1-G(C))^2}(Q(t))^{-1}.$$

In particular, if the offspring variance σ^2 is finite, then $E\nu(t) \sim e^C\sigma^2 t/2$ for $\theta(t) = \sigma^2 Ct/2$. If $\alpha = 1$ and $L(x) = -\log x$ in (2.1), then $E\nu(t) \sim e^C t \log t$ for $\theta(t) = Ct\log t$.

At last, in the supercritical case $(A > 1)$ if $EX(1)\ln X(1) < \infty$ and $\theta(t) = A^tC$ then as $t \to \infty$

$$E\nu(t) \sim (1-H(C))^{-1} \quad \text{and} \quad Var\nu(t) \sim H(C)(1-H(C))^{-2},$$

where the Laplace transform of $H(x)$ satisfies equation (2.4). In the case of positive geometric offspring from Example 2 and $\theta(t) = Cp^{-t}$ we have $E\nu(t) \to e^{Cp}$ and $Var\nu(t) \to e^{2pC}(1-e^{-Cp})$. In particular, when $p = 1/2$ and $C = 4$ the expected index of the process exceeding at time t level 2^{t+2} tends to e^2 as $t \to \infty$.

2.2 Sequence of processes with distinct offspring distributions

In this section we consider the sequence $\{X_i(t), i \in N\}$ of BGW processes under the assumption that each process has different offspring distribution. More precisely, we replace assumption (b) from Subsection 2.1 with

(b′) $P\{X_i(1) = k|X_i(0) = 1\} = p_k^{(i)}, \sum_{k=0}^{\infty} p_k^{(i)} = 1 \ (i = 1, 2, \ldots)$.

Denote

$$f^{(i)}(s) = \sum_{j=0}^{\infty} p_j^{(i)} s^j, \quad A_i = EX_i(1) \quad \text{and} \quad Q_i(t) = P\{X_i(t) > 0\}.$$

Again we start with the **subcritical case**. If $A_i < 1$, then

$$\lim_{t \to \infty} P\{X_i(t) = j|X_i(t) > 0\} = p_j^*(i) \quad (j \geq 1)$$

and the generating function $f^{(i)*}(s)$ of $\{p_j^*(i), (j \geq 1)\}$ satisfies the equation

$$1 - f^{(i)*}(f^{(i)}(s)) = A(1 - f^{(i)*}(s)).$$

Theorem 4. *If $A_i < 1$ and $\theta(t) \equiv \theta \in \{1, 2, \ldots\}$, then for any fixed k as $t \to \infty$*

$$P\{\nu(t) = k|X_k(t) > 0\} - q_k^*(\theta) = O\left(\sum_{i=1}^{k-1} q_i^*(\theta) Q_i(t)\right),$$

where

$$q_i^*(\theta) = \sum_{j=\theta+1}^{\infty} p_j^*(i).$$

Note that if $A_i \leq 1$, then $Q_i(t)$ tends to zero and therefore the theorem implies that for any fixed k

$$\lim_{t \to \infty} P\{\nu(t) = k|X_k > 0\} = q_k^*(\theta). \tag{2.5}$$

Under the additional condition $EX_i(1) \ln X_i(1) < \infty$ we can find the rate of convergence in (2.5). The following corollary holds.

Corollary 3. *If $A_i < 1$, $EX_i(1) \ln X_i(1) < \infty$, $(i \geq 1)$ and $\theta(t) \equiv \theta \in \{1, 2, \ldots\}$, then*

$$P\{\nu(t) = k\} A_k^{-1} - D_k q_k^*(\theta) = O(M^t),$$

where $M = \sup_i A_i < 1$ and

$$0 < D_i = \sum_{n=0}^{\infty} B_i(P\{X_i(n) = 0\}) < \infty \quad \textit{with} \quad B_i(s) = \frac{1 - f^{(i)}(s)}{A_i(1-s)}.$$

Next we present results for critical and supercritical processes, respectively. Recall that $m = \sup_{i \geq 1} A_i$.

Theorem 5. *If $A_i = 1$, $\sigma_i^2 = VarX_i(1) < \infty$ and $\theta(t) = Ct$ $(0 < C < \infty)$, then for any fixed k as $t \to \infty$*

$$P\{\nu(t) = k\}t - \frac{2}{\sigma_k^2} e^{-2C/\sigma_k^2} = O(t^{-1}).$$

Theorem 6. *If $A_i > 1$, $EX_i(1)\ln X_i(1) < \infty$ $(i \geq 1)$ and $\theta(t) = CM$ $(0 < C < \infty)$, then for any fixed k*

$$\lim_{t\to\infty} P\{\nu(t) = k\} = \begin{cases} (1 - H_k(C))\prod_j^* H_j(C) & \text{if } A_k = M \\ 0 & \text{if } A_k < M \end{cases},$$

where the product $\prod_j^*$ is over the values of $1 \leq j \leq k-1$ such that $A_j = M$ and $H_j(x)$ is the distribution function of the limiting variable $W_j = \lim_{t\to\infty} X_j(t)/A_j^t$.

The results in Subsections 2.1 and 2.2 are proved in [13]. In the next section we will continue our study of the process $\nu(t)$ in the critical case for a different class of level functions $\theta(t)$.

2.3 Extended class of level functions for critical processes

The proofs of the limit theorems for $\nu(t)$ in the previous section make use of some classical limit results for the population size. This imposes some restricting conditions on the level function $\theta(t)$. In this section we use a different technique based on a large deviation result for the critical case. This allows us to expand the class of level functions.

First, notice that if the offspring variance is finite, then the limit in (2.2) is exponential and thus for fixed $y \geq 0$

$$\lim_{t\to\infty} e^y P\{X(t)Q(t) > y | X(t) > 0\} = 1. \quad (2.6)$$

Let us introduce the following condition.

Condition A Assume that $A = 1$, $\sigma^2 = f''(1) < \infty$, and the function $f(s)$ is analytic in the disk $|s| < 1 + \varepsilon$ for some $\varepsilon > 0$.

It has been proved in [10] that if Condition A is satisfied, then (2.6) still holds when y is not fixed as long as for some natural $N \geq 2$,

$$0 < y = y(t) = o\left(\frac{t}{\log_{(N)} t \log t}\right) \quad (2.7)$$

as $t \to \infty$, where $\log_{(1)} t = \log t$, $\log_{(i+1)} t = \log\log_{(i)} t$, $i = 1, 2, \ldots$ Applying this result we obtain a limit theorem for $\nu(t)$ when the level function has the form $\phi(t)/Q(t)$ and $\phi(t)$ satisfies (2.7).

Theorem 7. *Let Condition* A *holds. If for some integer $N \geq 2$ we have*

$$\phi(t) = o\left(t/(\log_{(N)} t \log t)\right) \quad \text{and} \quad \theta(t) = \phi(t)/Q(t),$$

then $\nu(t) \in \mathcal{A}[\frac{y}{E\nu(t)}, E(x)]$, *where*

$$E\nu(t) \sim \frac{\sigma^2}{2} t e^{\phi(t)} \quad \textit{and} \quad E(x) = 1 - e^{-x}.$$

Proof. It is not difficult to obtain the following relation for the probability generating function of $\nu(t)$

$$Es^{\nu(t)} = \frac{s[1 - F(t, \theta(t))]}{1 - sF(t, \theta(t))} \quad (0 \le s \le 1), \tag{2.8}$$

where $F(t, \theta(t)) = P\{X(t) \le \theta(t)\}$. Let

$$a(t) = \frac{\sigma^2 t}{2} e^{\phi(t)}.$$

Setting in (2.8) $s = e^{-\lambda/a(t)}$ we have

$$\begin{aligned} Ee^{-\lambda\nu(t)/a(t)} &= \frac{e^{-\lambda/a(t)}\,[1 - F(t, \theta(t))]}{1 - e^{-\lambda/a(t)} + e^{-\lambda/a(t)}\,[1 - F(t, \theta(t))]} \qquad (2.9) \\ &= e^{-\lambda/a(t)} \frac{[1 - F(t, \theta(t))] / [1 - e^{-\lambda/a(t)}]}{1 + e^{-\lambda/a(t)}\,[1 - F(t, \theta(t))] / [1 - e^{-\lambda/a(t)}]}. \end{aligned}$$

If $\theta(t) = \phi(t)/Q(t)$, then using (2.2) we have

$$\begin{aligned} \frac{e^{\phi(t)}}{Q(t)} [1 - F(t, \theta(t))] &= e^{\phi(t)} P\{X(t)Q(t) > \phi(t) | X(t) > 0\} \\ &\to 1 \end{aligned} \tag{2.10}$$

On the other hand, since $Q(t) \sim 2/(\sigma^2 t)$ we have

$$\begin{aligned} \frac{e^{\phi(t)}}{Q(t)} \left[1 - e^{-\lambda/a(t)}\right] &= \frac{e^{\phi(t)}}{Q(t)} \frac{\lambda}{a(t)} (1 + o(1)) \\ &\to \lambda \end{aligned} \tag{2.11}$$

Relations (2.9)-(2.11) imply

$$\lim_{t\to\infty} Ee^{-\lambda\nu(t)/a(t)} = \frac{1}{1+\lambda} \quad (\lambda > 0)$$

and referring to the continuity theorem for Laplace transforms, we complete the proof. □

As an illustration of Theorem 7 let us take $\phi(t) = a\log t$, $(a > 0)$.

Corollary 4. *Let $f(s)$ be analytic on the disk $|s| < 1 + \varepsilon$ for some $\varepsilon > 0$. If $A = 1$, $0 < \sigma^2 < \infty$ and $\phi(t) = a\log t$, $(a > 0)$, then*

$$\lim_{t\to\infty} P\left\{\frac{2\nu(t)}{\sigma^2 t^{1+a}} \le x\right\} \to 1 - e^{-x} \quad (x \ge 0)$$

and $E\nu(t) \sim \sigma^2 t^{1+a}/2$ as $t \to \infty$.

2.4 Processes with immigration

In this section, instead of BGW processes we study processes with immigration. Assume that at the birth of the tth generation, that is, at time t, a number Y_t individuals immigrate into the population. The BGW process with immigration (BGWI) is defined by the recurrent relation, $Z(0) = 1$ and

$$Z(t) = \sum_{i=1}^{Z(t-1)} X_{ti} + Y_t,$$

where $Y_0, Y_1, \ldots$ are independent and identically distributed and independent of$\{X_{ti}\}$.

Let us consider the sequence of BGWI processes $\{Z_i(t), t \in N\}$ and define, as before, the process $\nu(t)$, $t \in N_0$ by

$$\nu(t) = \min\{i : Z_i(t) > \theta(t)\}$$

for a given function $\theta(t)$. We assume that $\{Z_i(t), i \in N\}$ are independent and identically distributed. This means that they have the same offspring distribution $\{p_k, k \geq 0\}$ and the distributions of the number of immigrants are also the same for all processes. Denote for $i = 1, 2, \ldots$

$$Z(t) \stackrel{d}{=} Z_i(t), \quad P\{Y_i = k\} = d_k \quad \text{and} \quad h(s) = \sum_{k=0}^{\infty} d_k s^k.$$

Next two results are for non-critical underline processes.

Theorem 8. *If $A < 1$, $0 < h'(1) < \infty$, and $\theta(t) \equiv \theta \in N = \{1, 2, \ldots\}$, then for any fixed $k \in N$,*

$$\lim_{t\to\infty} P\{\nu(t) = k\} = \left(1 - \sum_{i=1}^{\theta} \rho_i\right)\left(\sum_{i=1}^{\theta} \rho_i\right)^{k-1},$$

where $\{\rho_k, k \geq 0\}$ is a probability distribution with generating function $h(s)\prod_{t=1}^{\infty} h(f_t(s))$ and $f_t(s)$ is the t-fold composition of $f(s)$ with itself.

In the supercritical case, it is known (see [1], p.264) that if$A > 1$, then there exists a sequence of constants C_t such that $\{Z(t)/C_t\}$ converges with probability one to a random variable V. If $E\log Y_1 < \infty$, then $P\{V < \infty\} = 1$ and V has an absolutely continuous distribution on $(0, \infty)$; if $E\log Y_1 = \infty$ then $P\{V < \infty\} = 0$.

Theorem 9. *If $A > 1$ and $\theta(t) = C_t\theta$, $\theta \in (0, \infty)$ then*

$$\lim_{t\to\infty} P\{\nu(t) = k\} = [1 - R(\theta)][R(\theta)]^{k-1}$$

where $R(x) = P(V \leq x)$.

Now we consider the critical case. It has been proved in [12] that, if$A = 1$, $0 < f''(1) = \sigma^2 < \infty$, $h'(1) = h > 0$, then for fixed $x > 0$

$$S(x) = P\left\{\frac{2Z(t)}{\sigma^2 t} \leq x\right\} \to \frac{1}{\Gamma(\alpha)} \int_{-\infty}^{x} y^{\alpha-1} e^{-y} dy$$

as $t \to \infty$, where $\Gamma(\alpha)$ is the gamma function and $\alpha = 2h/\sigma^2$.

Theorem 10. *If $A=1$, $\sigma^2<\infty$, $h'(1)<\infty$, and $\theta(t)=\sigma^2 t\theta/2$, $\theta\in(0,\infty)$, then*

$$\lim_{t\to\infty} P\{\nu(t)=k\}=[1-S(\theta)][S(\theta)]^{k-1}.$$

Let the following condition be satisfied.

Condition B The generating functions $f(s)$ and $h(s)$ are analytic in the disk $|s|<1+\varepsilon$ for some $\varepsilon>0$, the maximal step of the lattice of the distribution of $f(s)$ is equal to 1 and $f'(1)=1$, $f''(1)=\sigma^2>0$, $h'(1)=h>0$.

Under Condition B, the asymptotic behavior of $P\{2Z(t)/\sigma^2 t>x\}$ was investigated in [11] when $t\to\infty$ and $x\to\infty$. The results of ([11]) allow us to prove a limit theorem for $\nu(t)$ when $\theta(t)$ tends to infinity faster than a linear function.

Theorem 11. *Assume Condition* B. *If $0<\phi(t)=o(t/\ln t)$ and $\theta(t)=\sigma^2 t\,\phi(t)/2$, then*

$$\nu(t)\in\mathcal{A}\left[\frac{y}{E\nu(t)},E(x)\right],$$

where $E\nu(t)\sim[1-S(\phi(t))]^{-1}$ and $E(x)=1-e^{-x}$.

For different choices of $\phi(t)$, Theorem 11 leads for different limit results for $\nu(t)$. Let, for example, the quantity $2h'(1)/\sigma^2$ be an integer. Then, using the asymptotic formula for the tail of the Gamma distribution

$$1-S(x)=e^{-x}\sum_{k=0}^{m}\frac{1}{k!}x^k\sim e^{-x}\frac{x^m}{m!}\qquad(x\to\infty),$$

we obtain the following result.

Corollary 5. *Assume Condition* B. *If $\phi(t)=\sqrt{t}/\ln t$ and $\theta(t)=\sigma^2 t\phi(t)/2$, then*

$$\lim_{t\to\infty} P\left\{\nu(t)\frac{e^{-\phi(t)}(\phi(t))^m}{m!}\le x\right\}\to 1-e^{-x}\qquad(x>0),$$

given that $m=2h'(1)/\sigma^2$ is an integer.

The results in Subsections 2.3 and 2.4 appeared in [14].

3 Number of Highly Productive Ancestors

In this section we assume a population with n ancestors. Each of these ancestors initiates a discrete time branching process. Let $\theta(t)$, $t\in N_0$ be a positive valued function and $S(n,t)$ be the number of ancestors having more than $\theta(t)$ descendants at time (generation) t. It is clear that if n is fixed and the process is critical or subcritical, then for any level function $\theta(t)$ and large enough t, $S(n,t)$ equals zero with probability one.

What is the asymptotic behavior of $S(n,t)$ when the number of ancestors increases to infinity? To answer this question we consider a family of stochastic processes $Y(x,t)=S([m(t)x],t)$, where $x\in[0,\infty)$ and $m(t)\to\infty$ as $t\to\infty$. We approximate $Y(x,t)$ by some known processes with independent increments. The behavior of the parameter $m(t)$ and the form of the limit processes naturally depend on the criticality of the underline branching process. It turns out that, if the branching process is supercritical, then $Y(x,t)$ may be approximated by either a "binomial process" (process with independent and binomially distributed increments) or by a Brownian motion depending on the behavior of $m(t)$. If the branching process is subcritical or critical, then the approximating process is either a Poisson process or a Brownian motion.

Now we give a rigorous definition of the process $S(n,t)$. Let $\mathcal{A}_t^i$ be the random population at time t generated by the i-th ancestor, $i = 1,2,\ldots,n$. For any positive valued function $\theta(t)$, the functional $S(n,t) = S(n,t)[\theta]$ can be defined as follows

$$S(n,t) = \#\{i: \text{ card } \mathcal{A}_t^i > \theta(t)\}.$$

Let $X_i(t) = \text{card } \mathcal{A}_t^i$ be the i-th branching process and $X(t)$ be a branching process such that $X(t) \stackrel{d}{=} X_i(t)$ for all $i \geq 1$. We denote $\{p_k, k \geq 0\}$ the offspring distribution of $X(t)$ and set as before

$$f(s) = \sum_{k=0}^{\infty} p_k s^k, \quad R(x,t) = P\{X(t) > x\}, \quad Q(t) = R(0,t),$$

$$A = \sum_{k=1}^{\infty} k p_k, \quad \sigma^2 = \sum_{k=1}^{\infty} k(k-1) p_k.$$

Let $D = D[0,1]$ be the space of functions $y(x)$ defined on $[0,1]$ that are right continuous and have left-hand limits at any point $x \in [0,1]$. It is clear that the realizations of $Y(x,t) = S([m(t)x],t)), x \in [0,1]$, are elements of D for any $t \in N_0$. We equip D with the Skorohod metric ([5], see also [2]). The topology of the weak convergence of measures on Borel sets in D induced by this metric is called the Skorohod topology. We approximate the process $Y(x,t)$ with respect to the Skorohod topology on $D[0,1]$. In the case of Poisson limiting process the convergence will be in the Skorohod topology on $D[0,\infty)$.
First we consider the critical case, i.e., the case of $A = 1,\ 0 < \sigma^2 < \infty$. Assume that

$$\lim_{t\to\infty} \frac{\theta(t)}{t} = \theta \in [0,\infty] \tag{3.1}$$

and consider $Y(x,t) = S([tx],t)$, i.e., $m(t) = t$.

Theorem 12. *If $A = 1,\ 0 < \sigma^2 < \infty$ and (3.1) is satisfied, then $Y(x,t) \xrightarrow{\mathcal{D}} Y(x)$ as $t \to \infty$, where $\mathcal{D}$ means convergence in the Skorohod topology on $D[0,\infty)$ and $Y(x)$, $x \in [0,\infty)$ is a Poisson process with $EY(x) = 2x\exp\{-2\theta/\sigma^2\}/\sigma^2$ for $\theta \in [0,\infty)$ and it is a "zero process" (i.e., $Y(x) \equiv 0$ with probability 1 for all $x \in [0,\infty)$) for $\theta = \infty$.*

The theorem gives an approximation of $S(n,t)$ for the case when $n = o(t)$ or $n \sim t$ as $t \to \infty$. Now we consider the case when $n/t \to \infty$ as $t \to \infty$. More precisely we set $m(t) = a(t)t$, where $a(t) \to \infty$. Define the stochastic process $W_t^{(1)}(x)$ as follows

$$W_t^{(1)}(x) = \frac{S([ta(t)x],t) - ta(t)xR(\theta(t),t)}{\sqrt{a(t)}},$$

where $R(\theta(t),t) = P\{X(t) > \theta(t)\}, x \in [0,1]$.

Theorem 13. *If* $A = 1$, $0 < \sigma^2 < \infty$ *and* (3.1) *is satisfied, then* $W_t^{(1)}(x) \xrightarrow{\mathcal{D}} W^{(1)}(x)$ *as* $t \to \infty$, *where* $\mathcal{D}$ *means convergence in the Skorohod topology on* $D[0,1]$ *and* $W^{(1)}(x)$ *is a Brownian motion with zero shift and diffusion parameter* $2\sigma^{-2}\exp\{-2\theta/\sigma^2\}$ *for* $\theta \in [0,\infty)$ *and it is a zero process for* $\theta = \infty$.

Example 3. Suppose that the ancestors are labeled by $1, 2, \ldots$ Let $\nu(t)$ be the number of the first ancestor having more than $\theta(t)$ descendants at time t, considered in Section 2. Then it follows from Theorem 12 that

$$\lim_{t\to\infty} P\{\nu(t) \le tx\} = \lim_{t\to\infty} P\{Y(x,t) \ge 1\} = P\{Y(x) \ge 1\}.$$

Thus we have the following limit theorem for $\nu(t)$.

Corollary 6. *If* $A = 1$, $0 < \sigma^2 < \infty$ *and condition* (3.1) *is satisfied, then* $\nu(t)/t \to \nu$ *as* $t \to \infty$ *in distribution and*

$$P\{\nu > x\} = \exp\left\{-\frac{2x}{\sigma^2} e^{-2\theta/\sigma^2}\right\}.$$

Consider the supercritical case. Recall that if $A > 1$, $EX(1)\ln X(1) < \infty$, then $X(t)A^{-t}$ converges with probability one to a random variable W and the Laplace transform $\varphi(\lambda)$ of W satisfies the equation (2.4). Let $H(x)$ be the distribution function of W. Assume that

$$\lim_{t\to\infty} \theta(t)A^{-t} = \theta \in [0,\infty] \quad \text{and} \quad \sum_{k=2}^{\infty} kP_k \ln k < \infty \tag{3.2}$$

and consider the "discrete time" process $S(n,t)$, $n = 0, 1, \ldots$ for $t \in N_0$. Note that here n is a time parameter.

Theorem 14. *If* $A > 1$ *and conditions* (3.2) *are satisfied, then* $S(n,t) \xrightarrow{w} \xi(n)$, $n \in N_0$ *as* $t \to \infty$, *where w means convergence in the weak sense and* $\xi(n)$ *is a stochastic process with independent and binomially distributed increments such that*

$$P\{\xi(n_i) - \xi(n_{i-1}) = k\} = \binom{n_i - n_{i-1}}{k} [1 - H(\theta)]^k H(\theta)^{n_i - n_{i-1} - k}$$

for any $0 \le n_{i-1} < n_i < \infty$, $n_i \in N_0$ *for* $\theta \in [0,\infty)$; *and it is a zero process for* $\theta = \infty$.

Example 4. Let the offspring distribution be positive geometric, i.e. $p_k = \alpha(1-\alpha)^{k-1}$, $k \ge 1$ and $p_0 = 0$. In this case the offspring generating function has the form $f(s) = \alpha s(1 - \beta s)^{-1}$, $\beta = 1 - \alpha$ and $A = \alpha^{-1}$ and the equation for the Laplace transform is:

$$\varphi\left(\frac{\lambda}{\alpha}\right) = \frac{\alpha\varphi(\lambda)}{1 - \beta\varphi(\lambda)}.$$

Now it is not difficult to check that the Laplace transform $\varphi(\lambda) = \alpha(\alpha+\lambda)^{-1}$ satisfies the above equation. Hence the limit distribution $H(x)$ is exponential with density function $\alpha e^{-\alpha x}$ and Theorem 14 gives the following result.

Corollary 7. *If the conditions of Theorem 14 are satisfied and the offspring distribution is positive geometric with parameter $0 < \alpha < 1$, then for $\theta \in [0, \infty)$ the limit process $\xi(n)$ in Theorem 14 is binomial such that*

$$P\{\xi(n_i) - \xi(n_{i-1}) = k\} = \binom{n_i - n_{i-1}}{k} e^{-\alpha\theta k} \left[1 - e^{-\alpha\theta}\right]^{n_i - n_{i-1} - k}.$$

Example 5. Consider the process $\nu(t)$ defined in Example 3. It follows from Theorem 14 that

$$\lim_{t \to \infty} P\{\nu(t) \leq n\} = \lim_{t \to \infty} P\{S(n,t) \geq 1\} = P\{\xi(n) \geq 1\}.$$

Consequently we obtain the following result, equivalent to Theorem 3.

Corollary 8. *If $A > 1$ and conditions (3.2) are satisfied, then for any fixed n*

$$\lim_{t \to \infty} P\{\nu(t) \leq n\} = 1 - H^n(\theta).$$

In particular, if the offspring distribution is as in Example 4, then the limit distribution is positive geometric with parameter $e^{-\alpha\theta}$.

The theorem above shows that the stochastic process $S(n,t)$ for fixed $n \in N_0$ can be approximated as $t \to \infty$ by a binomial process. Further we consider the case $A > 1$ and $n \to \infty$. Let $a(t)$ be a positive function such that $a(t) \to \infty$ as $t \to \infty$. Consider the following stochastic process

$$W_t^{(2)}(x) = \frac{S([a(t)x],t) - a(t)xR(\theta(t),t)}{\sqrt{a(t)}},$$

where $x \in [0,1]$.

Theorem 15. *If $A > 1$ and conditions (3.2) are satisfied, then $W_t^{(2)}(x) \xrightarrow{\mathcal{D}} W^{(2)}(x)$ as $t \to \infty$, where $W^{(2)}(x)$, $x \in [0,1]$, is a Brownian motion with zero shift and diffusion parameter $H(\theta)(1 - H(\theta))$ for $\theta \in [0, \infty)$; and it is a zero process for $\theta = \infty$.*

Example 6. If, as in Example 4, the offspring distribution is positive geometric of parameter $0 < \alpha < 1$, then it is not difficult to see that the Brownian motion in Theorem 15 has diffusion parameter $e^{-\alpha\theta}\left(1 - e^{-\alpha\theta}\right)$.

Let us turn to the case $A < 1$, i.e., the branching process is subcritical. Recall the notation for p_k^* and K from Theorem 1 in Section 2.

Theorem 16. *If $A < 1$ and $\sum kp_k \ln k < \infty$, then $Y(x,t) \xrightarrow{\mathcal{D}} Y(x)$ as $t \to \infty$, where $\mathcal{D}$ means convergence in Skorohod topology on $D[0, \infty)$ and $Y(x)$ is a Poisson process with*

$$EY(x) = Kx \sum_{j > \theta} p_j^*$$

for $\theta(t) \equiv \theta \in N_0$; and it is a zero process if $\theta(t) \to \infty$.

Consider the case $nA^t \to \infty$. Let, as before, $a(t)$ be a positive valued function such that $a(t) \to \infty$ as $t \to \infty$. Define the process $W_t^{(3)}(x)$ by the relation

$$W_t^{(3)}(x) = \frac{1}{\sqrt{a(t)}} \left\{ S([xA^{-t}a(t)],t) - xA^{-t}a(t)R(\theta(t),t) \right\},$$

where $x \in [0,1]$.

Theorem 17. *If $A < 1$ and $\sum kp_k \ln k < \infty$, then $W_t^{(3)}(x) \xrightarrow{\mathcal{D}} W^{(3)}(x)$ as $t \to \infty$, where $W^{(3)}(x)$, $x \in [0,1]$, is a Brownian motion with zero shift and diffusion parameter $K\sum_{i>\theta} p_i^*$ for $\theta(t) = \theta \in N_0$; and it is a zero process if $\theta(t) \to \infty$.*

The convergence of the finite dimensional distributions in Theorems 12–17 is proved in [15]. The proofs of the convergence in the Skorokhod topology are based on a representation of the process $S(n,t)$ as a sum of independent indicators and can be found in [16].

4 Large Families in Multitype Processes

In this section we consider a discrete time multitype Bienaym'e–Galton–Watson (MBGW) process. Each of the individuals of type i lives a unit time and at the end of its life independently of the others generates a random number of new individuals of types $1,2,\ldots,n$. These new individuals behave similarly. Suppose that the number of offspring of different individuals of the same type are independent and identically distributed random vectors with probability distribution $\{p_{\mathbf{k}}^{(i)}, \mathbf{k} \in N_0^n\}$, where $N_0^n = N_0 \times N_0 \times \cdots \times N_0$, $i = 1,2,\ldots,n$. In other words, $p_{\mathbf{k}}^{(i)}$, $\mathbf{k} = (k_1,\ldots,k_n)$ is the probability that a single individual of type i at the end of its life generates k_j individuals of type j, $j = 1,2,\ldots,n$. The process $\mathbf{X}(t) = (X_1(t),\ldots,X_n(t))$, $\mathbf{X}(0) = \delta_i = (\delta_{ij}, j = 1,\ldots,n)$, where $\delta_{ij} = 0$, $j \neq i$ and $\delta_{ii} = 1$, is called MBGW process initiated by a single ancestor of type i.

One variable related to genealogy of the process is the time to the closest common ancestor of the whole population at given time. For a single-type critical Galton–Watson process this, the time is asymptotically uniformly distributed (see [27]). The time to the closest common ancestor can be treated as a functional of the so called reduced branching processes. This process was introduced in [4] as a process that counts only those individuals at time τ having descendants at time t, $t > \tau$. In the critical single-type case the reduced process can be approximated by a non-homogeneous pure birth process. For extensions of these results to more general single and multitype branching processes see [25] and [26].

Here we study a random variable defined on a generalized reduced multitype process. Let $\theta = (\theta_1(t),\ldots,\theta_n(t))$ be a vector of non-negative functions, τ and $t, \tau < t$ be two time points. Let $\{i, i \in \nabla\}$, where $\nabla \subseteq \{1,2,\ldots,n\}$, denotes a subset of the types set. Define the process $\mathbf{X}(\tau,t) = (X_1(\tau,t),\ldots,X_n(\tau,t))$, where $X_i(\tau,t)$ is the number of type i individuals at time τ, whose number of descendants at time t of types j is greater than a threshold, given by a vector $\theta(t-\tau)$. If, for instance, $\nabla = \{1,2,\ldots,n\}$, then $X(\tau,t)$ is the number of individuals of type i at time τ whose the number of descendants at time t of all types are greater than a given threshold. It is clear that $\mathbf{X}(\tau,t)$ counts only "highly productive" individuals at time τ. There is an obvious one-to-one correspondence between the highly productive individuals at time τ and the large families at time t. Therefore, the components of the vector $\mathbf{X}(\tau,t)$ may be considered as the number of large families at time t. We will prove limit theorems for the process $\mathbf{X}(\tau,t)$ as $t,\tau \to \infty$ in the critical case, considering different cases of relationship between the time points τ and t.

Further on we will need the following notation. Let us have a single ancestor at time $t=0$ of type i. One direct descendant of the initial ancestor is denoted as (i,j,m_j), where j, $j=1,\ldots,n$ is the type of the direct descendant and $m_j \in N$, $N=\{1,2,\ldots\}$ is the label (number) of the descendant in the set of all immediate descendants of i. Thus m_{k+1}-th direct descendant of the type i_{k+1} of the individual $\alpha=(i,i_1,m_1,\ldots,i_k,m_k)$ will be denoted as $\alpha'=(\alpha,i_{k+1},m_{k+1})$. Further on for any two vectors $\alpha=(i_1,\ldots,i_k)$ and $\beta=(j_1,\ldots,j_m)$ the ordered pair (α,β) will be a $k+m$ dimensional vector $(i_1,\ldots,i_k,j_1,\ldots,j_m)$.

Making use of the above notation, it is clear that the set $\mathfrak{R}_n \in E$, where E is the space of all finite subsets of

$$\bigcup_{k=1}^{\infty} N_1^k, \quad N_1^k = N_1^{k-1} \times N_1, \quad N_1 = \{i\} \times \{1,\ldots,n\} \times N$$

corresponds to the population of the n-th generation. $\mathfrak{R}_n$ can be decomposed as $\mathfrak{R}_n = \bigcup_{i=1}^{n} \mathfrak{R}_n^{(i)}$, where $\mathfrak{R}_n^{(i)}$ is the population of the type i individuals of the n-th generation. Consequently, the components of the process $\mathbf{X}(t)$ are found as $X_i(t) = card\{\mathfrak{R}_n^{(i)}\}$, $t \in N_0$ and for any τ and t such that $\tau < t$ we have

$$\mathbf{X}(t) = \sum_{i=1}^{n} \sum_{\alpha \in \mathfrak{R}_\tau^{(i)}} \mathbf{X}^{(\alpha)}(t-\tau),$$

where $\mathbf{X}^{(\alpha)}(t) = (X_1^{(\alpha)}(t),\ldots,X_n^{(\alpha)}(t))$ is the n-type branching process generated by the individual α.

Let $\mathfrak{I}_i([\theta],\tau,t)$ be the set of individuals in $\mathfrak{R}_\tau^{(i)}$ having number of descendants at time t of types j, $j \in \nabla$ greater than the corresponding component of $\theta(t-\tau)$. It is not difficult to see that it can be described as

$$\mathfrak{I}_i([\theta],\tau,t) =$$
$$= \left\{\alpha \in \mathfrak{R}_\tau^{(i)} : \text{for all } j \in \nabla \ \exists \text{ more than } \theta_j(t-\tau) \ \beta\text{-sets such that } (\alpha,\beta) \in \mathfrak{R}_t^{(i)}\right\},$$

where $\alpha \in N_1^\tau$, $\beta \in N_1^{t-\tau}$. Thus the generalized reduced process is defined as $\mathbf{X}(\tau,t) = (X_i(\tau,t),\ i=1,\ldots,n)$ with $X_i(\tau,t) = \text{card}\{\mathfrak{I}_i([\theta],\tau,t)\}$.

In particular, if $\theta(t) = \mathbf{0}$ for all t, then $\mathfrak{I}_i([\mathbf{0}],\tau,t)$ contains all and only those individuals of type i of the τ-th generation having descendants (at least of one type) in generations $\tau+1,\tau+2,\ldots,t-1$ and in t-th generation having descendants of all types j, $j \in \nabla$ (recall that the initial n-type process is indecomposable). For instance, if $\nabla = \{1\}$, then $\mathbf{X}(\tau,t)$ counts all individuals of the τ-th generation having at least one descendant of type 1 in the t-th generation. If $\nabla = \{1,2,\ldots,n\}$, then it counts individuals of τ-th generation having descendants of all types in t-th generation.

Another process counting the large families in a population can be defined as the number of individuals at time τ of a given type whose number of descendants of at least one type at time t is greater than the corresponding threshold (see [17]). Note that the standard reduced process is a particular case of this process when the threshold is zero.

Let us introduce some additional notation. For n -dimensional vectors $\mathbf{x} = (x_1,\dots,x_n)$, $\mathbf{y} = (y_1,\dots,y_n)$ we denote $\mathbf{x}\otimes\mathbf{y} = (x_1y_1,\dots,x_ny_n)$, $\mathbf{x}^{\mathbf{y}} = (x_1^{y_1},\dots,x_n^{y_n})$, $\mathbf{x}/\mathbf{y} = (\frac{x_1}{y_1},\dots,\frac{x_n}{y_n})$, $(\mathbf{x},\mathbf{y}) = x_1y_1+\cdots+x_ny_n$, $\mathbf{1} = (1,1,\dots,1)$, $\mathbf{0} = (0,0,\dots,0)$, $\mathbf{e} = (e,e,\dots,e)$ and $\mathbf{x}\geq\mathbf{y}$ or $\mathbf{x}>\mathbf{y}$ if $x_i\geq y_i$ or $x_i>y_i$ for all i, respectively. Denote by p^i_α, $\alpha = (\alpha_1,\dots,\alpha_n)\in N_0^n$, the offspring distribution of the process $\mathbf{X}(t)$, i.e.

$$p^i_\alpha = P\{\mathbf{X}(1) = \alpha|\mathbf{X}(0) = \delta_i\}$$

is the probability that an individual of type i generates the total number α new individuals. Here $\delta_i = (\delta_{ij}, j = 1,\dots,n)$, $\delta_{ij} = 0$ if $i\neq j$ and $\delta_{ii} = 1$. Also denote

$$F^i(\mathbf{S}) = \sum_{\alpha\in N_0^n} p^i_\alpha S_1^{\alpha_1}\cdots S_n^{\alpha_n},\quad \mathbf{F}(\mathbf{S}) = (F^1(\mathbf{S}),\dots,F^n(\mathbf{S})),$$

$$Q^i(t) = P\{\mathbf{X}(t)\neq\mathbf{0}|\mathbf{X}(0) = \delta_i\},\quad \mathbf{Q}(t) = (Q^1(t),\dots,Q^n(t)),$$

i.e. $Q^i(t)$ is the probability of non-extinction up to time t for the process starting with a singly individual of type i.

Let for $i,j,k = 1,2,\dots,n$

$$a_i^j = \left.\frac{\partial F^j(\mathbf{S})}{\partial S_i}\right|_{\mathbf{S}=\mathbf{1}},\quad b_{ik}^j = \left.\frac{\partial^2 F^j(\mathbf{S})}{\partial S_i\partial S_k}\right|_{\mathbf{S}=\mathbf{1}},$$

and let $\mathbf{A} = \|a_i^j\|$ be the matrix of expectations with Peron root ρ and right and left eigenvectors $\mathbf{U} = (u_1,u_2,\dots,u_n)$ and $\mathbf{V} = (v_1,v_2,\dots,v_n)$ such that $\mathbf{AU} = \rho\mathbf{U}$, $\mathbf{VA} = \rho\mathbf{V}$, $(\mathbf{U},\mathbf{V}) = 1$, $(\mathbf{U},\mathbf{1}) = 1$.

If $\mathbf{A}$ is indecomposable, aperiodic and $\rho = 1$, then the process $\mathbf{X}(t)$ is called critical indecomposable multitype branching process. We assume that the generating function $\mathbf{F}(\mathbf{S})$ satisfies the following

$$x - \sum_{j=1}^{n} v_j\big(\mathbf{1} - F^j(\mathbf{1}-\mathbf{U}x)\big) = x^{1+\alpha}L(x), \tag{4.1}$$

where $0<x\leq 1, \alpha\in(0,1]$, and $L(x)$ is a slowly varying function as $x\downarrow 0$. Note that under this assumption $\rho = 1$, i.e., the process is critical and the second moments of the offspring distribution b_{ik}^j, $i,j,k = 1,\dots,n$, may not be finite. The following limit theorem for the process $\mathbf{X}(t)$ holds (see [23]).

Proposition 1 ([23]) *If the offspring generating function* $\mathbf{F}(\mathbf{S})$ *satisfies (4.1), then we have*

a) $Q^i(t)\sim u_j t^{-1/\alpha}L_1(t)$ *as* $t\to\infty$, *where* $L_1(t)$ *is a slowly varying as* $t\to\infty$;

b) $\lim_{t\to\infty} P\{\mathbf{X}(t)q(t)\leq\mathbf{x}\otimes\mathbf{V}\mid\mathbf{X}(t)\neq\mathbf{0},\mathbf{X}(0) = \delta_i\} = \pi(\mathbf{x})$, *where* $q(t) = \sum_{j=1}^{n} v_jQ^j(t)$ *and* $\pi(\mathbf{x}) = \pi(x_1,x_2,\dots,x_n)$ *is a distribution with Laplace transform*

$$\phi(\lambda) = \int_{R_+^n} e^{-(\mathbf{x},\lambda)}d\pi(\mathbf{x}) = 1-(1+\bar{\lambda}^{-\alpha})^{-1/\alpha},\quad \bar{\lambda} = (\lambda,\mathbf{1}). \tag{4.2}$$

where $\lambda = (\lambda_1,\dots,\lambda_n)$, $\lambda\in R_+^n$, $R_+ = [0,\infty)$.

Let us introduce a mapping $\Delta_{i_1\cdots_m}: R^n \mapsto R^n$, $m = 1,2,\ldots,n$ such that for each $\mathbf{x} = (x_1,\ldots,x_n)$ the vector $\Delta_{i_1,\ldots,i_m}(\mathbf{x})$ has $i_1,i_2,\ldots,i_m$-th coordinates as that of $\mathbf{x}$ and the remaining coordinates are ∞. For instance, $\Delta_{13}(\mathbf{x}) = (x_1,\infty,x_3,\infty,\ldots,\infty)$. Denote

$$L = \Sigma'\pi(\Delta_l(\theta)) - \Sigma'_{l\neq m}\pi(\Delta_{lm}(\theta)) + \cdots + (-1)^{|\nabla|-1}\pi(\theta),$$

where $|\nabla|$ is the cardinality of ∇ and Σ' is summation over all values of l and m from ∇, $\theta = (\theta_1,\ldots,\theta_n)$.

Let us state our first result about $\mathbf{X}(\tau,t)$. Let $\theta = (\theta_1,\ldots,\theta_n) \in R^n_+$ and $\mathbf{C} = (C_1,\ldots,C_n) \in R^n_+$ be some nonnegative vectors.

Theorem 18. *If condition* (4.1) *is satisfied,* $\theta(t) = \theta \otimes \mathbf{V}/q(t)$ *and* $t,\tau \to \infty$, $t-\tau \to \infty$ *such that* $\mathbf{Q}(t-\tau)/\mathbf{Q}(\tau) \to \mathbf{C}$, *then*

$$P\{\mathbf{X}(\tau,t) = \mathbf{k} | \mathbf{X}(\tau) \neq \mathbf{0}, \mathbf{X}(0) = \delta_i\} \to p^*_{\mathbf{k}},$$

where $\mathbf{k} = (k_1,\ldots,k_n) \in N^n_0$. *The probability distribution* $\{p^*_{\mathbf{k}}, \mathbf{k} \in N^n_0\}$ *has generating function* $\phi^*(\mathbf{S}) = \phi(\mathbf{a})$, *where* $\mathbf{a} = (1-L)\mathbf{C} \otimes \mathbf{U} \otimes \mathbf{V} \otimes (\mathbf{1}-\mathbf{S})$, $\mathbf{S} = (S_1,\ldots,S_n)$.

Remark It is clear that the vector $\mathbf{C}$ in the condition $\mathbf{Q}(t-\tau)/\mathbf{Q}(\tau) \to \mathbf{C}$ has the form $\mathbf{C} = C\mathbf{1}$, where $C \geq 0$ is some constant.

Example 7. Let $\mathbf{F}(\mathbf{S})$ satisfies condition (4.1) with $\alpha = 1$. Note that the second moments of the offspring distribution may still be infinite. For this process the limit distribution $\pi(\theta)$ is exponential and the generating function $\phi^*(\mathbf{S})$ has the form $\phi^*(\mathbf{S}) = (1+d)^{-1}$, where $d = (1-L)C\sum_{j=1}^{n} u_j v_j (1-S_j)$, and $\pi(\Delta_{i_1\cdots i_m}(\theta)) = 1 - \exp\{\min\{\theta_{i_1},\ldots,\theta_{i_m}\}\}$. Therefore,

$$L = \Sigma'(1-e^{-\theta_l}) - \Sigma'_{l\neq m}(1-e^{-\min\{\theta_l,\theta_m\}}) + \cdots + (-1)^{|\nabla|-1}(1-e^{-\min\{\theta_i, i\in\nabla\}}).$$

The generating function can be written as

$$\phi^*(\mathbf{S}) = \frac{1}{1+C(1-L)}\left(1 - \frac{C(1-L)}{1+C(1-L)}\sum_{i=1}^{n} u_i v_i S_i\right)^{-1}. \tag{4.3}$$

What is the distribution with this probability generating function? To answer this question we consider a sequence of independent random variables $X_1, X_2, \ldots$ such that $P\{X_i = j\} = p_j$, $j = 0,1,2,\ldots,n$, $\sum_{j=0}^{n} p_j = 1$, where $p_0 = (1+C(1-L))^{-1}$, $p_j = C(1-L)u_j v_j/(1+C(1-L))$, $j = 1,2,\ldots,n$. Let Δ_1 be the number of $1's$, Δ_2 be the number of $2'$s and so on, Δ_n be the number of n's observed in the sequence $X_1, X_2, \ldots$. before the first zero is obtained. It follows from the formula for the generating function of generalized multivariate geometric distribution in Ch. 36.9 of [8] that the vector $(\Delta_1,\ldots,\Delta_n)$ has the probability generating function (4.1), i.e.,

$$E\left(S_1^{\Delta_1} S_2^{\Delta_2} \cdots S_n^{\Delta_n}\right) = \phi^*(\mathbf{S}).$$

Hence we have the following result.

Corollary 9. *Under the assumptions of Theorem* 18 *and* $\alpha = 1$*, the limiting probability distribution* $\{p^*_{\mathbf{k}}, \mathbf{k} \in N_0^n\}$ *is multivariate geometric distribution with generating function* (4.1) *such that*

$$p^*_{\mathbf{k}} = P\{\Delta_i = k_i, i = 1, \ldots, n\}.$$

It is clear that, if $n = 1$, the limiting distribution is geometric, i.e., $p^*_k = pq^k$, $k = 0, 1, \ldots$ with $p = (1 + Ce^{-\theta_1})^{-1}$ and $q = Ce^{-\theta_1}(1 + Ce^{-\theta_1})^{-1}$.

Example 8. Let the assumptions of Theorem 18 be satisfied and $\tau = [\varepsilon t]$, $0 < \varepsilon < 1$. Referring to the asymptotic behavior of $\mathbf{Q}(t)$ and the uniform convergence theorem for slowly varying functions, we obtain as $t \to \infty$

$$\frac{\mathbf{Q}(t-\tau)}{\mathbf{Q}(\tau)} \to \left(\frac{\varepsilon}{1+\varepsilon}\right)^{1/\alpha} \mathbf{1}.$$

Consequently, $C = (\varepsilon/(1+\varepsilon))^{1/\alpha}$. We have the following result.

Corollary 10. *If the assumptions of Theorem* 18 *are satisfied and* $\tau = o(t)$*, then*

$$\lim_{t \to \infty} P\{\mathbf{X}(\tau, t) = \mathbf{k} | \mathbf{X}(\tau) \neq \mathbf{0}, \mathbf{X}(0) = \delta_i\} = 0$$

for all $\mathbf{k} \in N_0^n$ *such that* $\mathbf{k} \neq \mathbf{0}$.

In the critical case $\mathbf{X}(t)$ goes to extinction with probability 1. Corollary 9 shows that, if $\tau = o(t)$, even the conditioned process $\mathbf{X}(\tau, t)$ given $\mathbf{X}(\tau) \neq \mathbf{0}$ vanishes with probability approaching 1.

The previous theorem establishes the limit distribution of $\mathbf{X}(\tau, t)$ when the time points $\tau \to \infty$ and $t \to \infty$ are such that $\mathbf{Q}(t-\tau)/\mathbf{Q}(\tau)$ has a finite limit. Next we consider the case when this limit is not finite. Let $T_i(\tau, t) = Q^i(t-\tau)/Q^i(\tau)$ and $\mathbf{T}(\tau, t) = (T_1(\tau, t), \ldots, T_n(\tau, t))$.

Theorem 19. *If condition* (4.1) *holds,* $\theta(t) = \theta \otimes \mathbf{V}/q(t)$ *and* $T_i(\tau, t) \to \infty$ $(i = 1, 2, \ldots, n)$*, as* $t, \tau \to \infty$*, and* $t - \tau \to \infty$*, then*

$$P\left\{\frac{\mathbf{X}(\tau, \mathbf{t})}{\mathbf{T}(\tau, \mathbf{t})} \leq \mathbf{x} | \mathbf{X}(\tau) \neq \mathbf{0}, \mathbf{X}(0) = \delta_i\right\} \to \pi\left(\frac{1}{1-L}\mathbf{x}\right),$$

where $\pi(\mathbf{x})$, $\mathbf{x} \in R^n_+$ *is the distribution from Proposition* 1 *and* L *is defined just before Theorem* 18.

Remark It follows from the asymptotic behavior of $Q^i(t)$ that if $T_i(\tau, t) \to \infty$ for at least one i, then it holds for every $i = 1, 2, \ldots, n$.

Example 9. If matrix $\mathbf{A}$ is indecomposable, aperiodic, $\rho = 1$ and $b^i_{jk} < \infty$ $(i, j, k = 1, \ldots, n)$, then (4.1) is satisfied with $\alpha = 1$, $L(x) \to const$ as $x \to 0$. In this case, $Q^i(t) \sim 2u_i/\sigma^2 t$, $i = 1, \ldots, n$ as $t \to \infty$, where $\sigma^2 = \sum_{j,m,k=1}^{n} v_j b^j_{mk} u_m u_k$. Consequently, as $t \to \infty$

$$q(t) = \sum_{j=1}^{n} Q^j(t) v_j \sim \frac{2}{\sigma^2 t}$$

and $\theta(t) \sim \sigma^2 t \theta \otimes \mathbf{V}/2$. On the other hand, $T_j(\tau, t) \sim \tau/(t-\tau)$, $j = 1, \ldots, n$. Thus, $T_j(\tau, t) \to \infty$. If $\tau \sim t$ we obtain the following result from Theorem 19.

Corollary 11. *If* $\rho = 1$, $0 < \sigma^2 < \infty$, *and* $t, \tau \to \infty$, $t - \tau \to \infty$ *such that* $\tau \sim t$, *then*

$$P\left\{\frac{t-\tau}{\tau}\mathbf{X}(\tau,t) \le \mathbf{x} | \mathbf{X}(\tau) \ne \mathbf{0}, \mathbf{X}(0) = \delta_i\right\} \to 1 - \exp\left\{-\frac{x^*}{1-L}\right\},$$

where $\mathbf{x} \in R^n_+$, $x^* = \min\{x_1, \ldots, x_n\}$, *and* L *is as in Example* 7.

The last two theorems describe the asymptotic behavior of$\mathbf{X}(\tau,t)$ when $t - \tau \to \infty$. Let us consider the case $\tau = t - t_0$, where $t_0 \in (0,\infty)$ is a constant.

Theorem 20. *If the condition* (4.1) *is satisfied,* $t, \tau \to \infty$, $t - \tau = t_0 \in (0,\infty)$, *and* $\theta(t) \equiv \theta = (\theta_1, \ldots, \theta_n) \in R^n_+$, *then*

$$P\{\mathbf{X}(\tau,t) \otimes \mathbf{Q}(\tau) \le \mathbf{x} | \mathbf{X}(\tau) \ne \mathbf{0}, \mathbf{X}(0) = \delta_i\} \to \pi\left(\frac{\mathbf{x}}{\mathbf{R}(t_0)}\right),$$

where $\mathbf{x} \in R^n_+$ *and*

$$\mathbf{R}(t) = (R^1(t), \ldots, R^n(t)), R^i(t) = P\left\{\bigcap_{j=1}^n \{X_j(t) > \theta_j\} | \mathbf{X}(0) = \delta_i\right\}.$$

Remark It follows from Proposition 1 that

$$\frac{Q^i(\tau)}{Q^i(t)} \sim \left(\frac{t}{t-t_0}\right)^{1/\alpha} \frac{L_1(t-t_0)}{L_1(t)}$$

which shows that $Q^i(\tau) \sim Q^i(t)$ as $t, \tau \to \infty$, $t - \tau = t_0$ for each $i = 1, \ldots, n$. Therefore, after normalizing the vector functions, $\mathbf{Q}(\tau)$ in Theorem 20 can be replaced by $\mathbf{Q}(t)$.

The proofs of Theorems 18–20 are based on a representation of the process under consideration in the form of a vector of sums of a random number of independent indicators. First we prove limit theorems for the vector of random sums and then apply these theorems to obtain results for the branching process. The proofs can be found in [18].

Acknowledgments

My sincere thanks to King Fahd University of Petroleum and Minerals, Dhahran, Saudi Arabia, for all the support and facilities I had.

References

[1] Athreya K. E.; Ney P. E. *Branching Processes*. Springer-Verlag: New York, 1972.

[2] Billingsley, P. *Convegence of Probability Measures*. Wiley: New York, 1968.

[3] Feller W. *An Introduction to Probabbility Theory and Its Applications*. Wiley: New York, 1971, Vol. 2.

[4] Fleischmann, K. Siegmund-Schultze, R. The structure of reduced critical Galton-Watson processes, *Math. Nachr.* 1977, *79*, 233–241.

[5] Gikhman, I. I.; Skorohod, A. V. *The Theory of Stochastic Processes.* Springer-Verlag: New York, 1975, Vol. 1.

[6] Heathcote C. R. A branching process allowing immigration. *JRSS, Ser. B* 1965, *27*, 138–143.

[7] Jagers P. *Branching Processes with Biological Applications.* Wiley: London, 1975.

[8] Johnson, N. L.; Kotz, S.; Balakrishnan, N. *Discrete Multivariate Distributions.* Wiley: New York, 1997.

[9] Leadbetter M.; Lindgren G.; Rootzen H. *Extremes and related properties of random sequences and processes*. Springer: New York, 1983.

[10] Makarov G. D. Large deviations for a critical Galton-Watson process. *Theory of Probability and Its Applications* 1980, *25*, 3:481–492.

[11] Makarov G. D.; Large Deviation for Branching Processes with Immigration. *Mathematical Notes* 1982, *32*, 3:401–410.

[12] Pakes A. G. On the Critical Galton-Watson Process with Immigration. *J. Australian Mathematical Society* 1971, *12*, 476-482.

[13] Rahimov I.; Hasan, H. Exceedance problems concerning a family of branching processes, *Pak. J. Statist.* 1998, *14*, 1:37–47.

[14] Rahimov I.; Hasan H. Limit theorems for the exceedances of a sequence of branching processes, *Bulletin of the Malaysian Math. Society (Second Series)* 1998, *21*, 37–46.

[15] Rahimov I.; Hasan H. Limit theorems for the process of exceedances in large populations, *Bulletin of the Malaysian Math. Society (Second Series)* 2000, *23*, 1–8.

[16] Rahimov I. Approximation of exceedance processes in large populations. *Stochastic Models* 2001, *17*, 2:147–156.

[17] Rahimov I.; Muttlak H. Random Sums of Random Vectors and Multi Type Families of Productive Individuals. *International J. Matematics and Matematical Sciences* 2004, *19*, 975–990.

[18] Rahimov I. Limit Theorems for the size of Subpopulation of Productive Individuals. *Stochastic Models* 2004, *20*, 3:261–280.

[19] Robert F. S. *Applied Combinatorics.* Printice-Hall, NJ, 1984.

[20] Seneta E., An Explicit-limit Theorem for the critical Galton-Watson Process with Immigration. *JRSS* 1970, *32*, 149–152.

[21] Sevastyanov, B. A. *Branching Processes*. Nauka: Moscow, 1971 (in Russian).

[22] Slack R. S. A Branching Process with Mean One and Possibly Infinite Variance, *Probability theory and related fields* 1967, *9*, 139–143.

[23] Vatutin V. A. Limit theorems for critical Markov branching processes with several types of particles and infinite second moments. *Math. USSR- Sbornik* 1977, *32*, 215–225.

[24] Watts V.; Rootzen H.; Leadbetter M. Convergence in distribution of intermediate order statistics from stationary sequences. *Ann. Probab.* 1982, *10*, 653–662.

[25] Yakymov, A. L. Reduced branching processes. *Theory Probab. Appl.* 1980, *25*, 593–596.

[26] Yakymov, A. L. Asymptotic properties of subcritical and supercritical reduced branching processes. *Theory Probab. Appl.* 1985, *30*, 183–188.

[27] Zubkov, A. M. Limit distribution of the distance up to closest common ancestor, *Theory Probab. Appl.* 1975, *20*, 614–623.

In: Records and Branching Processes ISBN 978-1-60456-526-3
Editors: M. Ahsanullah, G. P. Yanev, pp. 127-145

Chapter 6

A REVIEW OF OFFSPRING EXTREMES IN BRANCHING PROCESSES

George P. Yanev*
Department of Mathematics, University of Texas - Pan American,
1201 West University Drive, Edinburg, Texas 78539, USA

Abstract

We summarize and discuss results on offspring extremes in branching processes that appeared in the last ten years.

Keywords and phrases: Branching processes, varying environments, bisexual processes, geometric arrays, maximum family sizes.

1 Introduction

There is a significant amount of research in the theory of branching processes devoted to extreme value problems concerning different population characteristics. The history of such studies goes back to the works in the fifties by Zolotarev [28] and Urbanik [24] (see also [7]) who considered the maximum generation size. Our goal here is to summarize and discuss results on maxima related to the offspring. Papers directly addressing this area of study have begun to appear in the last ten years (though see "hero mothers" example in [8].)

Let $\mathcal{M}_n$ denote the maximum offspring size of all individuals living in the $(n-1)$-st generation of a branching process. This is a maximum of random number of independent and identically distributed (i.i.d.) integer-valued random variables, where the random index is the population size of the process. $\mathcal{M}_n$ has two characteristic features: (i) the i.i.d. random variables are integer-valued and (ii) the distribution of the random index is connected to the distribution of the terms involved through the branching mechanism. These two characteristics distinguish the subject matter maxima among those studied in the general extreme value theory.

*E-mail address: gyanev@utpa.edu

The study of the sequence $\{\mathcal{M}_n\}$ might be motivated in different ways. It provides a fertility measure characterizing the most prolific individual in one generation. It also measures the maximum litter (or family) size. In the branching tree context, it is the maximum degree of a vertex. The asymptotic behavior of $\mathcal{M}_n$ gives us some information about the influence of the largest families on the size and survival of the entire population.

The chapter is organized as follows. Next section deals with maxima in simple branching processes with or without immigration. In Section 3 we review results about maxima of a triangular array of zero-modified geometric variables. Later we apply these to branching processes with varying geometric environments. An interesting parallel with some results for maxima of random number i.i.d. random variables is pointed out. Section 4 begins with limit theorems for the max-domain of attraction of bivariate geometric variables. Then we discuss one application to branching processes with promiscuous matting. The final section considers a different construction in which a random score (a continuous random variable) is associated with each individual in a simple branching process. We present briefly limiting results for the score's order statistics. In the end of the section, we give an extension to two-type processes.

2 Maximum Family Size in Simple Branching Processes

Define a Bienaymé–Galton–Watson (BGW) branching process and its n-th generation maximum family size by $Z_0 = 1$;

$$Z_n = \sum_{i=1}^{Z_{n-1}} X_i(n) \quad \text{and} \quad \mathcal{M}_n = \bigvee_{i=1}^{Z_{n-1}} X_i(n) \quad (n = 1,2,\ldots),$$

respectively, where the offspring variables $X_i(n)$ are i.i.d. nonnegative and integer-valued. Here and further on $\bigvee_{i=1}^{n} \xi_i = \max\{\xi_1, \xi_2, \ldots, \xi_n\}$ where ξ_i $(i \geq 1)$ are random variables.

Along with the BGW process $\{Z_n\}$, we consider the process with immigration $\{Z_n^{im}\}$ and its offspring maximum

$$Z_n^{im} = \sum_{i=1}^{Z_{n-1}^{im}} X_i(n) + Y_n \quad \text{and} \quad \mathcal{M}_n^{im} = \bigvee_{i=1}^{Z_{n-1}^{im}} X_i(n) \quad (n = 1,2,\ldots),$$

respectively, where $\{Y_n,\ n = 1,2,\ldots\}$ are independent of the offspring variables, i.i.d. and integer-valued non-negative random variables.

Finally, let us modify the immigration component such that immigrants may enter the nth generation only if the $(n-1)$st generation size is zero. Thus, we have the Foster-Pakes process and its offspring maximum

$$Z_n^0 = \sum_{i=1}^{Z_{n-1}^0} X_i(n) + I_{\{Z_{n-1}^0 = 0\}} Y_n \quad \text{and} \quad \mathcal{M}_n^0 = \bigvee_{i=1}^{Z_{n-1}^0} X_i(n) \quad (n = 1,2,\ldots),$$

where I_A stands for the indicator of A.

Denote by $F(x) = P(X_i(n) \leq x)$ the common distribution function of the offspring variables with mean $0 < m < \infty$ and variance $0 < \sigma^2 \leq \infty$. In this section, we consider the subcritical $(m < 1)$, critical $(m = 1)$, and supercritical $(m > 1)$ processes separately.

2.1 Subcritical Processes

Let $\widehat{\mathcal{M}_n}$ denote the maximum family size in all three processes defined above: $\{Z_n\}$, $\{Z_n^{im}\}$, and $\{Z_n^0\}$. Let $g(s)$ be the immigration p.g.f.. Also, let $\mathcal{A}_n = \{Z_{n-1} > 0\}$ for processes without immigration, and $\mathcal{A}_n$ be the certain event - otherwise. The following result is true.

Theorem 1. *If* $0 < m < 1$*, then for* $x \geq 0$

$$\lim_{n\to\infty} P(\widehat{\mathcal{M}_n} \leq x | \mathcal{A}_n) = \gamma(F(x)) \tag{2.1}$$

and

$$\lim_{n\to\infty} E(\widehat{\mathcal{M}_n} | \mathcal{A}_n) = \sum_{k=0}^{\infty} [1 - \gamma(F(k))] \tag{2.2}$$

where

(i) *in case of* $\{Z_n\}$*,* γ *is the unique p.g.f. solution of* $\gamma(f(s)) = m\gamma(s) + 1 - m$ *and* (2.2) *holds if, in addition,* $EX_i(n)\log(1 + X_i(n)) < \infty$*;*

(ii) *in case of process* $\{Z_n^{im}\}$*,* (2.1) *holds provided* $E\log(1 + Y_n) < \infty$ *and* γ *is the unique p.g.f. solution of* $\gamma(s) = g(s)\gamma(f(s))$*.* (2.2) *is true if, in addition,* $EY_n < \infty$*;*

(iii) *in case of process* $\{Z_n^0\}$ *we assume that* $E\log(1 + Y_n) < \infty$*. Then* $\gamma(s) = 1 - \sum_{n=0}^{\infty}[1 - g(f_n(s))]$ $(0 < s \leq 1)$ *and* $\gamma(0) = \{1 + \sum_{n=0}^{\infty}[1 - g(f_n(0))]\}^{-1}$*. Also,* (2.2) *holds if, in addition,* $EY_n < \infty$*.*

Example 1. Consider $\{Z_n\}$ with geometric offspring p.g.f. $f(s) = p/(1 - qs)$, where $1/2 < p = 1 - q < 1$. Then $m = q/p < 1$ and it is not difficult to see that $\gamma(s) = (1 - m)s/(1 - ms)$. Hence

$$\lim_{n\to\infty} P(\mathcal{M}_n \leq k \mid Z_{n-1} > 0) = \frac{(p - q)(1 - q^{k+1})}{p - q(1 - q^{k+1})}.$$

It can also be seen ([21]) that

$$\frac{m}{1 - pm} \leq \lim_{n\to\infty} E(\mathcal{M}_n | Z_n > 0) \leq \frac{m}{1 - m}.$$

Example 2. Consider $\{Z_n^{im}\}$ (see [16]) with

$$f(s) = (1 + m - ms)^{-1} \quad (0 < m < 1) \quad \text{and} \quad g(s) = f^{\nu}(s) \qquad (\nu > 0).$$

Then $\gamma(s) = ((1 - m)/(1 - ms))^{\nu}$, a negative binomial p.g.f., and the above theorem yields

$$\lim_{n\to\infty} P(\mathcal{M}_n^{im} \leq x) = \left(\frac{1 - m}{1 - mF(x)}\right)^{\nu} \text{ and } \lim_{n\to\infty} E\mathcal{M}_n^{im} = \sum_{j=0}^{\infty} 1 - \left[\frac{1 - F(j)}{1 - mF(j)}\right]^{\nu} \leq \frac{\nu m^2}{1 - m}.$$

Example 3. Let $\mu = EY_n$. Consider $\{Z_n^0\}$ with

$$f(s) = (1 + m - ms)^{-1} \qquad \text{and} \qquad g(s) = 1 - (\mu/m)\log(1 + m - ms) \qquad (0 < m < 1).$$

In this case $\gamma = (m - \mu\log(1 - ms))/(m - \mu\log(1 - m))$ and by the theorem

$$\lim_{n\to\infty} P\{\mathcal{M}_n^0 \leq x\} = \frac{m - \mu\log(1 - mF(x))}{m - \mu\log(1 - m)},$$

and

$$\lim_{n\to\infty} E\mathcal{M}_n^0 = \mu\,\frac{m+\sum_{k=0}^{\infty}\log\dfrac{1-m[(1+m)^{k+1}-m^{k+1}]}{1-m}}{m-\mu\log(1-m)} \le \frac{\mu m}{m-\mu\log(1-m)}\,\frac{m}{1-m}.$$

2.2 Critical Processes

In the rest of this section we need some asymptotic results for the maxima of i.i.d. random variables. Recall that a distribution function $F(x)$ belongs to the max-domain of attraction of a distribution function $H(x,\theta)$ (i.e., $F \in D(H)$) if and only if there exist sequences $a(n) > 0$ and $b(n)$ such that

$$\lim_{n\to\infty} F^n(a(n)x+b(n)) = H(x,\theta)\ , \tag{2.3}$$

weakly. According to the classical Gnedenko's result, $H(x;\theta)$ has the following (von Mises) form

$$H(x;\theta) = \exp\{-h(x;\theta)\} = \exp\left\{-(1+x\theta^{-1})^{-\theta}\right\}, \quad 1+x\theta^{-1} > 0;\ -\infty < \theta < \infty. \tag{2.4}$$

Necessary and sufficient conditions for $F \in D(H)$ are well-known. In particular, $F \in D(\exp\{-x^{-a}\})$, $a > 0$ if and only if for $x > 0$ the following regularity condition on the tail probability holds

$$1-F(x) = x^{-a}L(x)\ , \tag{2.5}$$

where $L(x)$ is a slowly varying at infinity function (s.v.f.).

A. Processes without immigration. In case of a simple BGW process, the following result holds.

Theorem 2. *Let* $m = 1$ *and* $0 < \sigma^2 < \infty$.

(i) *If* (2.3) *holds, then*

$$\lim_{n\to\infty} P\left(\frac{\mathcal{M}_n - b(n)}{a(n)} \le x|Z_{n-1} > 0\right) = \frac{1}{1+\sigma^2 h(x,\theta)/2}. \tag{2.6}$$

(ii) *If* (2.5) *holds, then*

$$\lim_{n\to\infty} \frac{E(\mathcal{M}_n|Z_{n-1} > 0)}{n^{1/a}L_1(n)} = \frac{\pi/a}{\sin(\pi/a)} \qquad (a \ge 2), \tag{2.7}$$

where $L_1(x)$ *is certain s.v.f. with known asymptotics.*

The theorem implies that if $F \in D(\exp\{-e^{-x}\})$ then the limiting distribution is logistic with c.d.f. $(1+e^{-x})^{-1}$; and if $F \in D(\exp\{-x^{-a}\})$ then the limiting distribution is log-logistic with c.d.f. $(1+x^{-a})^{-1}$.

Theorem 3. *Let $m = 1$, $\sigma^2 = \infty$, and (2.5) holds. Then for $x \geq 0$ and $1 < a \leq 2$*

$$\lim_{n\to\infty} P\left(\frac{\mathcal{M}_n}{n^{1/[a(a-1)]}L_2(n)} \leq x \mid Z_{n-1} > 0\right) = 1 - \frac{1}{\left(1+x^{a(a-1)}\right)^{1/(a-1)}}, \tag{2.8}$$

which is a Burr Type XII distribution (e.g. [23]) and

$$\lim_{n\to\infty} \frac{E(\mathcal{M}_n|Z_{n-1}>0)}{n^{1/[a(a-1)]}L_2(n)} = \frac{1}{a-1}B\left(\frac{1}{a-1} - \frac{1}{a(a-1)}, 1+\frac{1}{a(a-1)}\right) \quad (1 < a \leq 2), \tag{2.9}$$

where $B(u,v)$ is the Beta function and $L_2(x)$ is certain s.v.f. with known asymptotics.

Note that for $a = 2$ the right-hand sides in (2.6) (under assumption (2.5)) and (2.7) coincide with those in (2.8) and (2.9), respectively. The right-hand side in (2.9) is the expected value of the limit in (2.8) (see [23]).

Example 4. Let $1 - F(x) \sim x^{-2}\log x$. In this case one can check (see [21]) that Theorem 3 with $a = 2$ implies

$$\lim_{n\to\infty} P\left(\frac{\mathcal{M}_n}{n^{1/2}(\log n)^{3/2}} \leq x \mid Z_{n-1} > 0\right) = \frac{4x^2}{1+4x^2}.$$

for $x \geq 0$ and

$$\lim_{n\to\infty} \frac{E(\mathcal{M}_n|Z_{n-1}>0)}{n^{1/2}(\log n)^{3/2}} = \frac{\pi}{2}.$$

B. Processes with immigration $\{Z_n^{im}\}$. Let $\mu = EY_n$. We have the following theorem.

Theorem 4. *Assume that $m = 1$, $0 < \sigma^2 < \infty$, and $0 < \mu < \infty$.* (i) *If* (2.3) *holds, then*

$$\lim_{n\to\infty} P\left(\frac{\mathcal{M}_n^{im} - b(n)}{a(n)} \leq x\right) = \frac{1}{(1+\sigma^2 h(x,\theta)/2)^{2\mu/\sigma^2}}. \tag{2.10}$$

(ii) *If* (2.5) *is true, then*

$$\lim_{n\to\infty} \frac{E\mathcal{M}_n^{im}}{n^{1/a}L_2(n)} = \frac{2\mu}{\sigma^2}B\left(\frac{2\mu}{\sigma^2}+\frac{1}{a}, 1-\frac{1}{a}\right) \quad (a \geq 2), \tag{2.11}$$

where $B(u,v)$ is the Beta function and $L_2(x)$ is certain s.v.f. with known asymptotics.

The theorem implies that if $F \in D(\exp\{-e^{-x}\})$ then the limiting distribution is generalized logistic with c.d.f. $\left(1+\sigma^2 e^{-x}/2\right)^{-2\mu/\sigma^2}$; if $F \in D(\exp\{-x^{-a}\})$ then the limiting distribution is a Burr Type III (e.g. [23]) with c.d.f. $\left(1+\sigma^2 x^{-a}/2\right)^{-2\mu/\sigma^2}$. The right-hand side in (2.11) is the expected value of the limit in (2.10) (see [23]).

Theorem 5. *Let $m = 1$, $\sigma^2 = \infty$, and (2.5) holds. In addition, suppose*

$$\Theta(x) := -\int_0^x \log[1 - P(Z_t^{im} > 0)]dt = c\log x + d + \varepsilon(x), \tag{2.12}$$

where $\lim_{x\to\infty}\varepsilon(x) = 0$, $c > 0$, and d are constants. Then for $x \geq 0$,

$$\lim_{n\to\infty} P\left(\frac{\mathcal{M}_n^{im}}{n^{1/[a(a-1)]}L_2(n)} \leq x\right) = \frac{1}{(1 + x^{-a(a-1)})^c} \quad (1 < a \leq 2), \tag{2.13}$$

which is a Burr Type III distribution (e.g. [23]) and

$$\lim_{n\to\infty} \frac{E\mathcal{M}_n^{im}}{n^{1/[a(a-1)]}L_2(n)} = cB\left(c + \frac{1}{a(a-1)}, 1 - \frac{1}{a(a-1)}\right) \quad (1 < a \leq 2), \tag{2.14}$$

where $B(u,v)$ is the Beta function and $L_2(x)$ is certain s.v.f. with known asymptotics. The right-hand side in (2.14) is the expected value of the limit in (2.13).

Note that for $c = 1$ and $a = 2$ the right-hand sides in (2.13) and (2.14) coincide with those in (2.8) and (2.9), respectively. The condition (2.12) holds even when the immigration mean is not finite. Next example illustrates this point.

Example 5. Following [17], we consider offspring and immigrants generated by

$$f(s) = 1 - (1-s)(1 + (a-1)(1-s))^{-1/(a-1)} \text{ and } g(s) = \exp\{-\lambda(1-s)^{a-1}\},$$

respectively. Then (2.5) holds and (2.12) yields

$$\Theta(t) = (\lambda/(a-1))[\log t + \log(a-1) + \log(1 + (a-1)t)^{-1}].$$

Therefore,

$$\lim_{n\to\infty} P\left(\frac{\mathcal{M}_n^{im}}{n^{1/[a(a-1)]}L_3(n)} \leq x\right) = \frac{1}{(1 + x^{-a(a-1)})^{\lambda/(a-1)}} \quad (1 < a \leq 2),$$

and

$$\lim_{n\to\infty} \frac{E\mathcal{M}_n^{im}}{n^{1/[a(a-1)]}L_3(n)} = \frac{\lambda}{a-1}B\left(\frac{\lambda}{a-1} - \frac{1}{a(a-1)}, 1 - \frac{1}{a(a-1)}\right) \quad (1 < a \leq 2),$$

where $B(u,v)$ is the Beta function and $L_3(x)$ is certain s.v.f. with known asymptotics.

C. Foster-Pakes processes $\{Z_n^0\}$. The following limit theorem for $\mathcal{M}_n^0$ under a non-linear normalization holds.

Theorem 6. *Assume that $m = 1$, $0 < \sigma^2 < \infty$, and $0 < \mu < \infty$. If*

$$\lim_{n\to\infty} \frac{P(X_1(1) > n)}{P(X_1(1) > n+1)} = 1 \tag{2.15}$$

then for $0 < x < 1$,

$$\lim_{n\to\infty} P\left(\frac{\log U(\mathcal{M}_n^0)}{\log n} \leq x\right) = x, \tag{2.16}$$

where $U(y) = 1/(1 - F(y))$.

Note that (2.15) is a necessary condition for $X_1(n)$ to be in a max-domain of attraction.

2.3 Supercritical Processes

Denote by $\widehat{\mathcal{M}_n}$ (as in the subcritical case above) the maximum family size in all three processes: $\{Z_n\}$, $\{Z_n^{im}\}$, and $\{Z_n^0\}$. The following result is true.

Theorem 7. *Assume that* $m > 1$ *and* $EX_i(n)\log(1+X_i(n)) < \infty$. *If* (2.3) *holds, then*

$$\lim_{n\to\infty} P\left(\frac{\widehat{\mathcal{M}_n} - b(m^n)}{a(m^n)} \leq x\right) = \psi(h(x,\theta))$$

If (2.5) *is true, then*

$$\lim_{n\to\infty} \frac{E\widehat{\mathcal{M}_n}}{m^{-n/a}L_1\left(m^{-n/a}\right)} = \int_0^\infty 1 - \psi(x^{-a})dx,$$

where $L_1(x)$ *is certain s.v.f. with known asymptotics.*

(i) *In case of* $\{Z_n\}$, ψ *is the unique, among the Laplace transforms, solution of*

$$\psi(u) = f(\psi(um^{-1})), \qquad (u > 0). \tag{2.17}$$

(ii) *In case of* $\{Z_n^{im}\}$, *we assume in addition that* $E\log(1+Y_n) < \infty$ *and then*

$$\psi(u) = \prod_{k=1}^{\infty} g(\varphi(um^{-k})) \qquad (u > 0)\ ,$$

where $\varphi(u)$ *is the unique, among the Laplace transforms, solution of*(2.17).

(iii) *In case of* $\{Z_n^0\}$, *we assume in addition that* $EY_n < \infty$ *and then*

$$\psi(u) = g(\varphi(u)) - \sum_{n=0}^{\infty}[1 - f(\varphi(um^{-n}))]P(Z_n^0 = 0) \qquad (u > 0)$$

where $\varphi(u)$ *is the unique, among the Laplace transforms, solution of*(2.17).

It is interesting to compare the limiting behavior of the maximum family size in the processes allowing immigration with that when the processes evolve in "isolation", i.e., without immigration. In the supercritical case, as might be expected, the immigration has little effect on the asymptotics of the maximum family size. The limits differ only in the form of the Laplace transform $\psi(u)$. In the subcritical and critical cases the mechanism of immigration eliminates the conditioning on non-extinction. Theorem 6 for the Foster-Pakes process differs from the rest of the results by the non-linear norming of $\mathcal{M}_n$. The study of the limiting behavior of the expectation in this case needs additional efforts.

It is known that some of the most popular discrete distributions, like geometric and Poisson, do not belong to any max-domain of attraction. This restricts the applicability of the results in the critical and supercritical cases above. A general construction of discrete distributions attracted in a max-domain is given in Wilms (1994). As it is proved there, if X is attracted by a Gumbel or Fréchet distributions, then the same holds for the integer part $\lfloor X \rfloor$. Next we follow a different approach considering triangular arrays of geometric variables which leads to branching processes with varying environments.

The results in this section are published in [10], [13], and [19]–[21]. In [27] an extension for order statistics is considered.

3 Maximum Family Size in Processes with Varying Environments

It is well-known that the geometric law is not attracted to any max-stable law. Therefore, the limit theorems for maxima in the critical and supercritical cases above do not apply to geometric offspring. In this section we utilize a triangular array of zero-modified geometric (ZMG) offspring distributions, instead.

3.1 Maxima of Arrays of Zero-Modified Geometric Variables

In this subsection we prove limit theorems for maximum of ZMG with p.m.f.

$$P(X_i(n) = j) = \begin{cases} a_n p_n (1-p_n)^{j-1} & \text{if } j \geq 1, \\ 1 - a_n & \text{if } j = 0, \end{cases} \qquad (n = 1, 2, \ldots)$$

For a positive integer ν_n consider the triangular array of variables

$$\begin{array}{llll} X_1(1), X_2(1), & \ldots & , X_{\nu_1}(1) & \\ X_1(2), X_2(2), & \ldots & , \quad . \; X_{\nu_2}(2) & \\ & \ldots & & \\ X_1(n), X_2(n), & \ldots & , & X_{\nu_n}(n) \end{array}$$

We prove limit theorems as $\nu_n \to \infty$ for the row maxima

$$\mathcal{M}_n = \bigvee_{i=1}^{\nu_n} X_i(n).$$

Let Λ have the standard Gumbel law with c.d.f. $\exp(-e^{-x})$ for $-\infty < x < \infty$.

Theorem 8. *Assume that for some real c*

$$\lim_{n\to\infty} p_n = 0 \quad \textit{and} \quad \lim_{n\to\infty} p_n \log(\nu_n a_n) = 2c.$$

A. If $\lim_{n\to\infty} \log(\nu_n a_n) = \infty$, *then* $c \geq 0$ *and*

$$p_n \mathcal{M}_n - \log(\nu_n a_n) \xrightarrow{d} \Lambda - c.$$

B. If $\lim_{n\to\infty} \log(\nu_n a_n) = \alpha$, $(-\infty < \alpha < \infty)$, *then*

$$p_n \mathcal{M}_n \xrightarrow{d} (\Lambda + \alpha)^+.$$

The idea of the proof is to exploit: (i) the exponential approximation to the zero-modified geometric law when its mean a_n/p_n is large; (ii) the fact that exponential law is attracted by Gumbel distribution.

3.2 Processes with Varying Geometric Environments

Consider a branching process with ZMG offspring law defined over the triangular array above. Thus, we have a simple branching process with geometric varying environments. For this process we prove limit theorems for the offspring maxima in all three classes: subcritical, critical, and supercritical. Define $\mu_0 = 1$,

$$\mu_n = E(Z_n|Z_0 = 1) = \prod_{j=1}^{n} m_j \qquad (n \geq 1).$$

If the environments are weakly varying, i.e., $\mu = \lim_{n\to\infty} \mu_n$ exists, then the processes can be classify (see [12]) as follows.

$$\{Z_n\} \text{ is } \begin{cases} \text{supercritical} & \text{if } \mu = \infty & \text{i.e. } \sum_n (m_n - 1) \to \infty \\ \text{critical} & \text{if } \mu \in (0,\infty) & \text{i.e. } \sum_n (m_n - 1) < \infty \\ \text{subcritical} & \text{if } \mu = 0 & \text{i.e. } \sum_n (m_n - 1) \to -\infty \end{cases}$$

Define the maximum family size for a process with varying geometric environments as

$$\mathcal{M}_n^{ge} = \bigvee_{i=1}^{Z_n} X_i(n), \qquad (n = 1,2,\ldots)$$

In the result below the role played by ν_n before is played by B_{n-1} where

$$B_n = \mu_n \sum_{j=1}^{n} \frac{p_j^{-1} - 1}{\mu_j}. \tag{3.1}$$

Let $\mathcal{V}$ be a standard logistic random variable with c.d.f. $(1+e^{-x})^{-1}$ for $-\infty < x < \infty$.

Theorem 9. *Suppose that* $\lim_{n\to\infty} B_n = \infty$ *and for c real*

$$\lim_{n\to\infty} p_n = 0 \quad \textit{and} \quad \lim_{n\to\infty} p_n \log(B_{n-1}a_n) = 2c.$$

A. If $\lim_{n\to\infty} \log(B_{n-1}a_n) = \infty$, *then*

$$(p_n \mathcal{M}_n^{ge} - \log(B_{n-1}a_n)|Z_{n-1} > 0) \xrightarrow{d} \mathcal{V} - c.$$

B. If $\lim_{n\to\infty} \log(B_{n-1}a_n) = \alpha$, $(-\infty < \alpha < \infty)$, *then*

$$(p_n \mathcal{M}_n^{ge}|Z_{n-1} > 0) \xrightarrow{d} (\mathcal{V} + \alpha)^{+}.$$

Referring to Theorems 8 and 9, we can say that the branching mechanism transforms Gumbel to logistic distribution. It is interesting to notice that this is in parallel with some results for maximum of i.i.d. random variables with random geometrically distributed index discussed in [6]. Let us make this more precise. Gnedenko and Gnedenko (1982) [6] have obtained the following result (see also [5]). Let $\xi_1, \xi_2, \ldots$ be i.i.d. random variables and there exist constants C_n and D_n such that

$$D_n \bigvee_{i=1}^{n} \xi_i - C_n \xrightarrow{d} \Lambda, \tag{3.2}$$

where, as before, Λ has the standard Gumbel law. Assume also that the random variable G_n is independent of $\{\xi_i\}$ and $G_n \sim Ge(1/n)$, i.e.,

$$P(G_n = k) = \left(1 - \frac{1}{n}\right)\left(\frac{1}{n}\right)^{k-1} \qquad k = 2, 3, \ldots$$

Then

$$D_n \bigvee_{i=1}^{G_n} \xi_i - C_n \xrightarrow{d} \mathcal{V}, \tag{3.3}$$

where, as before, $\mathcal{V}$ has the standard logistic law. Notice also that

$$\frac{G_n}{n} \xrightarrow{d} \mathcal{E}, \tag{3.4}$$

where $\mathcal{E}$ has the standard exponential law.

Now, let us look back at Theorem 8 and replace ν_n by $[B_n]$ from (3.1). Under the assumptions of Theorem 8, we have

$$\left(p_n \bigvee_{i=1}^{[B_n]} X_i(n) - \log(B_n a_n) |\, Z_{n-1} > 0\right) \xrightarrow{d} \Lambda - c \quad (c \geq 0), \tag{3.5}$$

where Λ has the standard Gumbel law. Theorem 9 implies that if (3.5) holds then

$$\left(p_n \bigvee_{i=1}^{Z_n} X_i(n) - \log(B_{n-1} a_n) |\, Z_{n-1} > 0\right) \xrightarrow{d} \mathcal{V} - c \quad (c \geq 0), \tag{3.6}$$

where $\mathcal{V}$ has the standard logistic law. Furthermore, it can be proved (see [12]) that if $\lim_{n\to\infty} B_n = \infty$, then

$$\left(\frac{Z_n}{B_n} |\, Z_n > 0\right) \xrightarrow{d} \mathcal{E}, \tag{3.7}$$

where $\mathcal{E}$ has the standard exponential law. Now we can say that, in case of geometric varying environments and under the assumptions of Theorems 8 and 9 above, the relations (3.2)-(3.4) are transformed to (3.5)-(3.7). I think this needs further investigations.

Example 6. Let us sample a linear birth and death process $(\mathcal{B}_t)$ at irregular times. Let $Z_n = \mathcal{B}_{t_n}$ where $0 < t_n < t_{n+1} \to t_\infty \leq \infty$. If λ and μ are the birth and death rates, respectively, and $d_n = t_n - t_{n-1}$, then $a_n = m_n p_n$,

$$p_n = \begin{cases} \dfrac{\lambda - \mu}{\lambda m_n - \mu} & if\ \lambda \neq \mu, \\ \dfrac{1}{1 + \lambda d_n} & if\ \lambda = \mu, \quad m_n = e^{(\lambda-\mu)d_n}. \end{cases}$$

and

$$B_n = \begin{cases} \frac{\lambda(\mu_n - 1)}{\lambda - \mu} & if\ \lambda \neq \mu, \\ \lambda t_n & if\ \lambda = \mu, \quad \mu_n = e^{(\lambda-\mu)t_n}. \end{cases}$$

A. If $\lambda > \mu$ and

$$\lim_{n\to\infty} \frac{t_n}{m_n} = \frac{2c}{\lambda - \mu} \in [0, \infty),$$

then

$$\left(\frac{\mathcal{M}_n^{ge}}{m_n} - (\lambda - \mu)t_n \mid Z_{n-1} > 0 \right) \xrightarrow{d} \mathcal{V} - c.$$

B. If $\lambda = \mu$ and $t_n = n^{\delta} l(n)$ $(\delta \geq 1)$ and l(n) - s.v.f., then

$$\left(\frac{\mathcal{M}_n^{ge}}{\lambda\delta n^{\delta-1} l(n)} - \log n \mid Z_{n-1} > 0 \right) \xrightarrow{d} \mathcal{V}.$$

The results in this section can be found in [12].

4 Maxima in Bisexual Processes

In this section we consider maxima of triangular arrays of bivariate geometric random vectors. The obtained results are applied to a class of bisexual branching processes.

4.1 Max-Domain of Attraction of Bivariate Geometric Arrays

The following construction is due to Marshall and Olkin [9]. Consider a random vector (U,V) having Bernoulli marginals, i.e., it takes on four possible values (0,0), (0,1), (1,0), and (1,1) with probabilities p_{00}, p_{01}, p_{10}, and p_{11}, respectively. Thus the marginal probabilities for U and V are

$$P(U=0) = p_{0+} = p_{00} + p_{01}, \qquad P(U=1) = p_{1+} = p_{10} + p_{11}$$
$$P(V=0) = p_{+0} = p_{00} + p_{10}, \qquad P(V=1) = p_{+1} = p_{01} + p_{11}.$$

Consider a sequence $\{(U_n, V_n)\}_{n=1}^{\infty}$ of independent and identically distributed with (U,V) random vectors. Let ξ and η be the number of zeros preceding the first 1 in the sequences $\{U_n\}_{n=1}^{\infty}$ and $\{V_m\}_{n=1}^{\infty}$, respectively. Both ξ and η follow a geometric distribution and, in general, they are dependent variables. The vector (ξ, η) has a bivariate geometric distribution with probability mass function for integer l and k

$$P(\xi = l, \eta = k) = \begin{cases} p_{00}^{l} p_{10} p_{+0}^{k-l-1} p_{+1} & \text{if } 0 \leq l < k, \\ p_{00}^{l} p_{11} & \text{if } l = k, \\ p_{00}^{k} p_{01} p_{0+}^{l-k-1} p_{1+} & \text{if } 0 \leq k < l, \end{cases} \tag{4.1}$$

and

$$P(\xi > l, \eta > k) = \begin{cases} p_{00}^{l+1} p_{+0}^{k-l} & \text{if } 0 \leq l \leq k, \\ p_{00}^{k+1} p_{0+}^{l-k} & \text{if } 0 \leq k < l. \end{cases} \tag{4.2}$$

The marginals of ξ and η for integer l and k are $P(\xi = l) = p_{1+}p_{0+}^{l} \quad (l \geq 0)$ and $P(\eta = k) = p_{+1}p_{+0}^{k} \quad (k \geq 0)$, respectively and

$$\bar{F}_{\xi}(l) = P(\xi > l) = p_{0+}^{l+1} \quad (l \geq 0), \qquad \bar{F}_{\eta}(k) = P(\eta > k) = p_{+0}^{k+1} \quad (k \geq 0). \tag{4.3}$$

For $n = 1, 2, \ldots$, let ν_n be a positive integer and $\{(\xi_i(n), \eta_i(n)) : i = 1, 2, \ldots, \nu_n\}$ be a triangular array of independent random vectors with the same bivariate geometric distribution (4.1) where p_{ij} are replaced by $p_{ij}(n) \ (i, j = 0, 1)$ for $n = 1, 2, \ldots$ That is,

$$\begin{array}{lll}
(\xi_1(1), \eta_1(1)), (\xi_2(1), \eta_2(1)), & \ldots, & (\xi_{\nu_1}(1), \eta_{\nu_1}(1)) \\
(\xi_1(2), \eta_1(2)), (\xi_2(2), \eta_2(2)), & \ldots, & (\xi_{\nu_2}(2), \eta_{\nu_2}(2)) \\
 & \ldots & \\
(\xi_1(n), \eta_1(n)), (\xi_2(n), \eta_2(n)), & \ldots, & (\xi_{\nu_n}(n), \eta_{\nu_n}(n))
\end{array}$$

Next we present a limit theorem for the bivariate row maximum as $\nu_n \to \infty$ (see [26] for the proof).

$$(\mathcal{M}_n^{\xi}, \mathcal{M}_n^{\eta}) = \left(\bigvee_{i=1}^{\nu_n} \xi_i(n), \bigvee_{i=1}^{\nu_n} \eta_i(n) \right).$$

Theorem 10. *Let $\lim_{n\to\infty} \nu_n = \infty$. If there are constants $0 \leq a, b, c < \infty$, such that*

$$\lim_{n\to\infty} p_{11}(n) \log \nu_n = 2c \quad \lim_{n\to\infty} \frac{p_{10}(n)}{p_{11}(n)} \log \nu_n = a \quad \textit{and} \quad \lim_{n\to\infty} \frac{p_{01}(n)}{p_{11}(n)} \log \nu_n = b,$$

then for $x, y \geq 0$

$$\begin{aligned}
&\lim_{n\to\infty} P\left(p_{11}(n)\mathcal{M}_n^{\xi} - \log \nu_n \leq x, \ \ p_{11}(n)\mathcal{M}_n^{\eta} - \log \nu_n \leq y\right) \\
&\quad = \exp\left\{-e^{-x-a-c} - e^{-y-b-c} + e^{-\max\{x,y\}-a-b-c}\right\}.
\end{aligned}$$

In particular, if $a = b = 0$ then

$$\lim_{n\to\infty} P\left(p_{11}(n)\mathcal{M}_n^{\xi} - \log \nu_n \leq x, p_{11}(n)\mathcal{M}_n^{\eta} - \log \nu_n \leq y\right) = \exp\left\{-e^{-\min\{x,y\}-c}\right\}.$$

Note that in this case the limit is proportional to the upper bound for the possible asymptotic distribution of a multivariate maximum given in [4], Theorem 5.4.1.

For the componentwise maxima, applying Theorem 10, one can obtain the following limiting results. If $p_{1+}(n) \log \nu_n \to 2c_1 < \infty$, then

$$\lim_{n\to\infty} P(p_{11}(n)\mathcal{M}_n^{\xi} - \log \nu_n \leq x) = \exp\left\{-e^{-x-c_1}\right\}.$$

If $p_{1+}(n) \log \nu_n \to 2c_2 < \infty$, then

$$\lim_{n\to\infty} P(p_{11}(n)\mathcal{M}_n^{\eta} - \log \nu_n \leq y) = \exp\left\{-e^{-y-c_2}\right\}.$$

4.2 Bisexual Processes with Varying Geometric Environments

Consider the array of bivariate random vectors $\{(\xi_i(n),\eta_i(n)) : i=1,2,\ldots;\ n=0,1,\ldots\}$, which are independent with respect to both indexes. Let $L : \mathcal{R}^+ \times \mathcal{R}^+ \to \mathcal{R}^+$ be a mating function. A bisexual process with varying environments is defined (see [15]) by the recurrence: $Z_0 = N > 0$,

$$(Z_{n+1}^F, Z_{n+1}^M) = \sum_{i=1}^{Z_n} (\xi_i(n),\eta_i(n)) \quad \text{and} \quad Z_{n+1} = L(Z_{n+1}^F, Z_{n+1}^M) \qquad n=0,1,\ldots$$

Define the mean growth rate per mating unit by

$$r_{nj} = j^{-1}E(Z_{n+1}|Z_n = j) \quad (j=1,2,\ldots) \quad \text{and} \quad \mu_n = \prod_{i=0}^{n-1} r_{i1},\ \mu_0 = 1 \quad (n=1,2,\ldots)$$

Further on we assume that $(\xi_i(n),\eta_i(n))$ are i.i.d. copies of the bivariate geometric vector (ξ,η) introduced above and that the mating is promiscuous, i.e.,

$$L(\xi(n),\eta(n)) = \xi(n)\min\{1,\eta(n)\}. \tag{4.4}$$

Theorem 11. *Let $\{Z_n\}$ be a bisexual branching process with varying geometric environments and mating function* (4.4). *If*

$$\prod_{j=1}^{\infty} p_{+0}(j)p_{0+}(j) \neq 0 \quad \textit{and} \quad \sum_{n=0}^{\infty} p_{+1}(n) < \infty\,,$$

then

$$\lim_{n\to\infty} \frac{Z_n}{\mu_n} = W \quad a.s.,$$

where W is a nonnegative random variable with $E(W) < \infty$ and $P(W > 0) > 0$.

Define offspring maxima in the bisexual process $\{Z_n\}$ by

$$(\mathcal{M}_n^F, \mathcal{M}_n^M) = \left(\bigvee_{i=1}^{Z_n} \xi_i(n),\ \bigvee_{i=1}^{Z_n} \eta_i(n)\right).$$

Theorem 12. *Assume that $\mu_n \to \infty$ and there are constants $0 \le a,b,c < \infty$, such that*

$$\lim_{n\to\infty} p_{11}(n)\log\mu_n = 2c \quad \lim_{n\to\infty} \frac{p_{10}(n)}{p_{11}(n)}\log\mu_n = a \quad \textit{and} \quad \lim_{n\to\infty} \frac{p_{01}(n)}{p_{11}(n)}\log\mu_n = b. \tag{4.5}$$

Also assume that

$$\prod_{j=1}^{\infty} p_{+0}(j)p_{0+}(j) \neq 0 \quad \textit{and} \quad \sum_{n=0}^{\infty} p_{+1}(n) < \infty\,. \tag{4.6}$$

Then

$$\lim_{n\to\infty} P\left(p_{11}(n)\mathcal{M}_n^F - \log\mu_n \le x, p_{11}(n)\mathcal{M}_n^M - \log\mu_n \le y\right) = \int_0^{\infty} (G(x,y))^z dP(W \le z),$$

where

$$G(x,y) = \exp\left\{-e^{-x-a-c} - e^{-y-b-c} + e^{-\max\{x,y\}-a-b-c}\right\}.$$

Theorems 11 and 12 are proved in [26]. Next example, adopted from [11], shows that the various conditions in Theorem 12 can be satisfied.

Example 7. Let $\alpha > 1$ and $\beta > 1$. Set

$$p_{11}(n) = n^{-\alpha} \qquad \text{and} \qquad p_{01}(n) = p_{10}(n) = n^{-(\alpha+\beta)} \qquad (n \geq 2).$$

It is not difficult to see that with this choice of $p_{ij}(n)$ $(i, j = 0, 1)$, we have

$$\log \mu_n \sim \alpha n \log n \qquad \text{as} \qquad n \to \infty$$

and both (4.5) (with $a = b = c = 0$) and (4.6) are satisfied.

5 Maximum Score

In this section we assume that every individual in a Galton-Watson family tree has a continuous random characteristic which maximum is of interest.

5.1 Maximum Scores in Galton-Watson Processes

Let us go back to the simple BGW process and attach a random score to each individual in the family tree. More specifically, associate with the j-th individual in the n-th generation a continuous random variable $Y_j(n)$. Arnold and Villaseñor (1996) published the first paper studying the maximum individual scores ("heights"). Pakes (1998) proves more general results concerning the laws of offspring score order statistics. Quoting [18], "these results provide examples of the behavior of extreme order statistics of observations from samples of random size." Define by $M_{(k),n}$ the k-th largest score within the n-th generation and by $\overline{M}_{(k),n}$ the k-th largest among the random variables $\{Y_i(n) :\ 1 \leq i \leq Z_\nu, 0 \leq \nu \leq n\}$, i.e., the k-th largest score up to and including the n-th generation. Pakes (1998) studies the limiting behavior of "near maxima", i.e., (upper) extreme order statistics $M_{(k),n}$ and $\overline{M}_{(k),n}$ when $n \to \infty$ and k remains fixed. The two general cases that arise are whether the law of Z_n (or the total progeny $T_n = \sum_{\nu=0}^{n} Z_\nu$), conditional on survival, does not require or does require, normalization to converge to a non-degenerate limit.

If no normalization is required then no particular restriction need to be placed on the score distribution function S, but the limit laws are rather complex mixtures of the laws of extreme order statistics. The principal result states that

$$\lim_{n\to\infty} P(M_{(k),n} \leq x | \mathcal{A}_n) = \sum_{j=1}^{\infty} \sum_{i=0}^{k-1} \binom{j}{i} (1 - S(x))^i S^{j-i}(x) g_j,$$

where it is assumed that the conditional law $\mathcal{G}_n$ of Z_n given $\mathcal{A}_n$ ($\mathcal{A}_n$ includes non-extinction) converges to a discrete and non-defective limit $\mathcal{G}$ and g_j denote the masses attributed to j by $\mathcal{G}$.

If normalization is required then one must assume that the score distribution function S is attracted to an extremal law, and then the limit laws are mixtures of the classical limiting laws of extreme order statistics. Let us assume that there are positive constants $C_n \uparrow \infty$ such that for the conditional law $\mathcal{G}_n$ we have $\mathcal{G}_n(xC_n) \Rightarrow N(x)$, where $N(x)$ is a non-defective but possibly degenerate distribution function. Assume also that the score distribution function S is in the domain of attraction of an extremal law given by (2.4). The general result in [18] is

$$\lim_{n\to\infty} P\left(\frac{M_{(k),n} - b(C_n)}{a(C_n)} \le x|\mathcal{A}_n\right) = \sum_{i=0}^{k-1} \frac{(h(x,\theta))^i}{i!} \int_0^\infty y^i e^{-yh(x,\theta)} dN(y). \tag{5.1}$$

Example 8. Consider an immortal (i.e., $P(X=0)=0$) supercritical process with shifted geometric offspring law given by its p.g.f. $f(s) = s/(1+m-ms)$ $(m>1)$, then (see Pakes (1998)) (5.1) becomes

$$\lim_{n\to\infty} P\left(\frac{M_{(k),n} - b(C_n)}{a(C_n)} \le x|\mathcal{A}_n\right) = 1 - \left(\frac{h(x,\theta)}{1+h(x,\theta)}\right)^k.$$

Thus the limit has a generalized logistic law when the score law is attracted to Gumbel law, $h(x,\theta) = e^{-x}$; and a Pareto-type law results when S is attracted to the Fréchet law.

Phatarford (see [18]) has raised the question (in the context of horse racing), "What is the probability that the founder of a family tree is better than all its descendants?" The answer turns out to be $E\left(T^{-1}\right)$, where $T = \sum_{n=0}^{\infty} Z_n$ is the total number of individuals in the family tree. More generally, if τ_n is the index of the generation up to the n-th which contains the largest score, Pakes (1998) proves that

$$P(\tau_n = k) = E\left(\frac{Z_k}{T_n}\right), \quad (k=0,1,\dots,n),$$

as well as limit theorems for τ_n as $n \to \infty$.

This subsection is based on [2] and [18].

5.2 Maximum Scores in Two-Type Processes

Let each individual in a two-type branching process be equipped with a non-negative continuous random variable - individual score. We present limit theorems for the maximum individual score. Consider two independent sets of independent random vectors with integer nonnegative components

$$\{\mathbf{X}^1(n)\} = \{(X^1_{1j}(n), X^1_{2j}(n))\} \text{ and } \{\mathbf{X}^2(n)\} = \{(X^2_{1j}(n), X^2_{2j}(n))\} \ (j \ge 1; n \ge 0).$$

A two-type branching process $\{\mathbf{Z}(n)\} = \{(Z_1(n), Z_2(n))\}$ is defined as follows: $\mathbf{Z}(0) \neq \mathbf{0}$ a.s. and for $n = 1, 2, \dots$

$$Z_1(n) = \sum_{j=1}^{Z_1(n-1)} X^1_{1j}(n) + \sum_{j=1}^{Z_2(n-1)} X^2_{1j}(n),$$

$$Z_2(n) = \sum_{j=1}^{Z_1(n-1)} X^1_{2j}(n) + \sum_{j=1}^{Z_2(n-1)} X^2_{2j}(n).$$

Here $X^i_{kj}(n)$ refers to the number of offspring of type k produced by the j-th individual of type i. With the j-th individual of type i living in the n-th generation we associate a non-negative continuous random variable $\zeta_{ij}(n)$, $(i=1,2)$ "score", say. Assume that the offspring of type 1 and type 2 have scores, which are independent and identically distributed within each type. Define the maximum score within the n-th generation by

$$\mathcal{M}_n^{\zeta} = \max\{\mathcal{M}_n^{\zeta_1},\ \mathcal{M}_n^{\zeta_2}\}, \qquad \text{where} \quad \mathcal{M}_n^{\zeta_i} = \max_{1\le j\le Z_i(n-1)} \zeta_{ij}(n) \qquad (i=1,2).$$

Note that this is maximum of random number, independent but non-identically distributed random variables. Let $F_i(x) = P(\zeta_i \le x)$ $(i=1,2)$ be the c.d.f.'s of the scores of type 1 and type 2 individuals, respectively.

Assumption 1 (tail-equivalence) We assume that F_1 and F_2 are tail equivalent, i.e., they have the same right endpoint x_0 and for some $A>0$

$$\lim_{x\uparrow x_0} \frac{1-F_1(x)}{1-F_2(x)} = A.$$

Assumption 2 (max-stability) Suppose F_1 is in a max-domain of attraction, i.e., (2.3) holds.

We consider the critical branching process $\mathbf{Z}(n)$ with mean matrix $\mathbf{M}$, which is positively regular and nonsingular. Let $\mathbf{M}$ have maximum eigenvalue 1 and associated right and left eigenvectors $\mathbf{u}=(u_1,u_2)$ and $\mathbf{v}=(v_1,v_2)$, normalized such that $\mathbf{u}\cdot\mathbf{v}=1$ and $\mathbf{u}\cdot\mathbf{1}=1$.

Theorem 13. *Let $\{\mathbf{Z}(n)\}$ be the above critical two-type branching process. If the offspring variance $2B<\infty$ and both Assumptions 1 and 2 hold, then*

$$\lim_{n\to\infty} P\left(\frac{\mathcal{M}_n^{\zeta} - b(v_1Bn)}{a(v_1Bn)} \le x \middle| \mathbf{Z}(n)\ne\mathbf{0}\right) = \frac{1}{1+h(x,\theta)+(v_2/v_1)h(cx+d,\theta)}, \tag{5.2}$$

where if $-\infty<\theta<\infty$ is fixed, then $c=A^{1/|\theta|}$ and $d=0$; if $\theta\to\pm\infty$, then $c=1$ and $d=\ln A$.

Proof. Since $F_1(x)$ and $F_2(x)$ are tail-equivalent, we have (see [22], p. 67)

$$\lim_{n\to\infty} (F_2(a(n)x+b(n)))^n \to H(cx+d,\theta),$$

where the constants c and d are as in (5.2). On the other hand, it is well-known (see [3], p. 191) that for $x>0$ and $y>0$

$$\lim_{n\to\infty} P\left(\frac{Z_1(n)}{v_1Bn}\le x,\ \frac{Z_2(n)}{v_2Bn}\le y \middle| \mathbf{Z}(n)\ne\mathbf{0}\right) = G(x,y),$$

where the limiting distribution has Laplace transform

$$\psi(\lambda,\mu) = \frac{1}{1+\lambda+\mu} \qquad (\lambda>0,\ \mu>0). \tag{5.3}$$

Set $x_n = a(v_1Bn)x + b(v_1Bn)$, $s_n = k/v_1Bn$, and $t_n = l/v_2Bn$. Referring to the definition of both $\mathcal{M}_n^{\zeta}$ and process $\{\mathbf{Z}(n)\}$ we obtain

$$P\left(\mathcal{M}_n^{\zeta} \le x_n | \mathbf{Z}_n \neq \mathbf{0}\right) = \sum_{(k,l)=\mathbf{0}}^{\infty} P(\mathbf{Z}(n) = (k,l)|\mathbf{Z}(n) \neq \mathbf{0}) P\left(\max\left\{\mathcal{M}_n^{\zeta_1}, \mathcal{M}_n^{\zeta_2}\right\} \le x_n\right)$$

$$= \sum_{(k,l)=\mathbf{0}}^{\infty} P\left(\frac{Z_1(n)}{v_1Bn} = \frac{k}{v_1Bn}, \frac{Z_2(n)}{v_2Bn} = \frac{l}{v_2Bn} |\mathbf{Z}(n) \neq \mathbf{0}\right) [F_1(x_n)]^k [F_2(x_n)]^l$$

$$= \sum_{(k,l)=\mathbf{0}}^{\infty} P\left(\frac{Z_1(n)}{v_1Bn} = s_n, \frac{Z_2(n)}{v_2Bn} = t_n|\mathbf{Z}(n) \neq \mathbf{0}\right) [F_1(x_n)]^{(v_1Bn)s_n} [F_2(x_n)]^{(v_1Bn)t_n(v_2/v_1)}$$

$$\to \int_0^{\infty}\int_0^{\infty} H(x,\theta)^s H(cx+d,\theta)^{(v_2/v_1)t} dG(s,t)$$

$$= \int_0^{\infty}\int_0^{\infty} \exp\left\{-sh(x,\theta) - t\frac{v_2}{v_1}h(cx+d,\theta)\right\} dG(s,t)$$

$$= \left[1 + h(x,\theta) + \frac{v_2}{v_1}h(cx+d,\theta)\right]^{-1},$$

where in the last formula we used the Laplace transform of $G(u,v)$ given in (5.3). The proof is complete. □

The two examples below illustrate the kind of limit laws that can be encountered.

Example 9. Let F_1 and F_2 be Pareto c.d.f.'s given for $x_i > \theta_i > 0$ and $c > 0$ by

$$F_i(x_i) = 1 - \left(\frac{\theta_i}{x_i}\right)^c \qquad (i = 1,2).$$

Note that the two distributions share the same value of the parameter c. It is not difficult to check that the limit is log-logistic given by

$$\lim_{n\to\infty} P\left\{\frac{\mathcal{M}_n^{\zeta}}{\theta_1(v_1Bn)^{1/c}} \le x|\mathbf{Z}(n) \neq \mathbf{0}\right\} = \left[1 + \left(1 + \frac{v_2}{v_1}\left(\frac{\theta_1}{\theta_2}\right)^{-c}\right)x^{-c}\right]^{-1}.$$

Example 10. Let F_1 and F_2 be logistic and exponential c.d.f.'s given by

$$F_1(x_1) = 1 - e^{-x_1} \quad (0 < x_1 < \infty) \quad \text{and} \quad F_2(x_2) = \frac{1}{1+e^{-x_2}} \quad (-\infty < x_2 < \infty),$$

respectively. It is known that both are in the max-domain of attraction of $H(x) = \exp\{-\exp\{-x\}\}$ and share (see [1], p. 91) the same normalizing constants $a(n) = 1$ and $b(n) = \ln n$. This fact, after inspecting the proof of the theorem, allows us to bypass the tail-equivalence assumption and obtain a logistic limiting distribution, i.e, for $-\infty < x < \infty$

$$\lim_{n\to\infty} P\left\{\mathcal{M}_n^{\zeta} - \log(v_1Bn) \le x \mid \mathbf{Z}(n) \neq \mathbf{0}\right\} = \left[1 + \left(1 + \frac{v_2}{v_1}\right)e^{-x}\right]^{-1}.$$

The results in this subsection are modifications of those in [14].

Acknowledgment

I thank I. Rahimov for igniting my interest to extreme value problems in branching processes.

References

[1] Ahsanullah, M.; Nevzorov, V. B., Ordered Random Variables. *Nova Science Publishers: Huntington, NY*, 2001.

[2] Arnold, B.C.; Villaseñor, J. A. The tallest man in the world. In Nagaraja, H. N.; Sen, P. K.; Morrison, D. F. (Eds). *Statist. Theory and Appl.: Papers in honor of Herbert A. David.* Springer: Berlin, 1996, 81–88.

[3] Athreya, K. B.; Ney, P. E. Branching Processes. *Springer: New York*, 1972.

[4] Galambos, J. The Asymptotic Theory of Extreme Order Statistics. *Krieger: Melbourne, FL*, 1987.

[5] Gnedenko, B. V. On some stability theorems. In *Lecture Notes in Math.* 982, Springer: Berlin-New York, 1983, 24–31.

[6] Gnedenko, B. V.; Gnedenko, D. B. Laplace distributions and the logistic distribution as limit distributions in probability theory. *Serdica*, 1982, *8*, 2:229–234 (in Russian).

[7] Harris, T. E. *The Theory of Branching Processes.* Springer: Berlin, 1963.

[8] Jagers, P.; Nerman, O. Limit theorems for sums determined by branching and exponentially growing processes. *Stoch. Proc. Appl.* 1984, *17*, 47–71.

[9] Marshall, A. W.; Olkin, I. A family of bivariate distributions generated by the Bernoulli distribution. *J. Amer. Statist. Assoc.* 1985, *80*, 332–338.

[10] Mitov, K. V. The maximal number of offspring of one particle in a branching process with state-dependent immigration. *Proc. of 27th Spring Conference of the Union of Bulgarian Mathematicians, Math. and Math. Education* 1998, 92–97.

[11] Mitov, K. V. Extremes of bivariate geometic variables with application to bisexual branching processes. *Pliska Studia Math. Bulgarica* 2005, *17*, 349-362.

[12] Mitov, K. V.; Pakes, A. G.; Yanev, G. P. Extremes of geometric variables with applications to branching processes. *Statist and Probab Letters* 2003, *65*, 379–388.

[13] Mitov, K. V.; Yanev G. P. Maximum family size in branching processes with state–dependent immigration. *Proc. of 28th Spring Conference of the Union of Bulgarian Mathematicians, Montana, Math. and Math. Education* 1999, 142–144.

[14] Mitov, K. V.; Yanev, G. P. Maximum individual score in critical two-type branching processes. *C. R. Acad. Bulg. Sci.* 2002, *55*, 11:17–22.

[15] Molina, M.; Mota, M.; Ramos, A. Limiting behaviour for superadditive bisexual Galton-Watson processes in varying environments. *Test* 2004, *13*, 2:481–499.

[16] Pakes, A. G. Branching processes with immigration. *J. Appl. Prob.* 1971, *8*, 32–42.

[17] Pakes, A. G. Some new limit theorems for the critical branching process allowing immigration. *Stoch. Proc. Appl.* 1975, *3*, 175–185.

[18] Pakes, A. G. Extreme order statistics on Galton–Watson trees. *Metrika* 1998, *47*, 95–117.

[19] Rahimov, I.; Yanev G. P. On a maximal sequence associated with simple branching processes. *Institute of Mathematics and Informatics, Sofia, Preprint* 1996, *6*, pp.14.

[20] Rahimov, I.; Yanev, G. P. Maximal number of direct offspring in simple branching processes. *Nonlinear Analysis, Theory, Methods and Applications* 1997, *30*, 2015–2023.

[21] Rahimov, I.; Yanev, G. P. On maximum family size in branching processes. *J. Appl. Probab.* 1999, *36*, 632–643.

[22] Resnick, S. *Extreme Value Distributions, Regular Variations, and Point Processes.* Springer: Berlin, 1987.

[23] Tadikamalla, P. R. A look at the Burr and related distributions. *International Statistical Review* 1980, *48*, 337–344.

[24] Urbanik, K. On a problem concerning birth and death processes. *Acta Math. Acad. Sci. Hungar.* 1956, *7*, 99–106 (in Russian).

[25] Wilms, R. Fractional parts of random variables. Limit theorems and infinite divisibility. *Ph. D. Thesis,* Technical University of Eindhoven, Eindhooven, Holland, 1994.

[26] Yanev, G. P. Revisiting offspring maxima in branching processes. *Pliska Studia Mathematica Bulgarica* 2007, *18*, 401–426.

[27] Yanev, G. P.; Tsokos, C. P. Family size order statistics in branching processes with immigration. *Stoch. Anal. Appl.* 2000, *18*, 4:655–670.

[28] Zolotarev, V. M. On a problem in the theory of branching rpocesses. *Uspehi Matemat. Nauk* 1954, *9*, 147–156 (in Russian).

In: Records and Branching Processes ISBN 978-1-60456-526-3
Editors: M. Ahsanullah, G. P. Yanev, pp. 147-172

Chapter 7

STATISTICAL INFERENCE FOR BRANCHING PROCESSES

Nikolay M. Yanev*
Department of Probability and Statistics, Institute of Mathematics and Informatics, Bulgarian Academy of Sciences, 8 Acad. G. Bonchev St., 1113 Sofia, Bulgaria

Abstract

We discuss non-parametric estimation for discrete-time single-type branching processes under different observation schemes. First we show that consistent and asymptotically normal estimators for the offspring characteristics of the classical Bienaymé-Galton-Watson (BGW) process exist only on the explosion set in the supercritical case. This restriction can be avoided introducing an increasing random number of ancestors or an immigration component. In the former case an unified estimation theory is developed for subcritical, critical and supercritical processes under different sample schemes. We also present the fundamental statistical results for branching processes with immigration. Then we show how to transfer asymptotic results for BGW processes with an increasing random number of ancestors to that for BGW processes with immigration designing a suitable sampling scheme. Finally, we discuss briefly some other directions of statistical inference for branching processes.

Keywords and phrases: Non-parametric estimators, limiting distributions, consistency and asymptotic normality, random number of ancestors, immigration.

1 Introduction

It is well known that the mathematical statistics considers the inverse problem of the probability theory. In the context of branching processes, given the individual characteristics the probability theory studies the global probabilistic behavior of the processes. In statistics we would like to estimate these unknown characteristics having real observations over the processes. Of course, in one chapter it is not possible to cover all different models of branching processes along with the existing schemes of observations.

*E-mail address: yanev@math.bas.bg

There is a number of publications which deal with statistical problems for branching processes and among them let us mention here some books and survey articles: Dion and Keiding (1978), Basawa and Scott (1983), Nanthi (1983), Iosifescu et al. (1984), Vatutin and Zubkov (1985), Winnicki (1988), Sankaranarayanan (1989), Badalbaev and Mukhitdinov (1990), Guttorp (1991), Dion (1993). Also recall that the background of the theory of branching processes is well presented in the monographs Harris (1963), Sevastyanov (1971), Mode (1971), Athreya and Ney (1972), Jagers (1975), Asmussen and Hering (1983). Some biological applications and estimation problems are considered in Jagers (1975), Yakovlev and Yanev (1989), Kimmel and Axelrod (2002), Haccou et al. (2005).

In this chapter we consider some estimation problems for Bienaymé-Galton-Watson (BGW) branching processes, with or without immigration. First we show in Section 2 that consistent and asymptotically normal estimators exist only on the explosion set for supercritical BGW processes. It is possible to avoid this restriction by considering either an immigration component or an increasing random number of ancestors. In the later we develop in Section 3 an unified estimation theory for subcritical, critical and supercritical processes using different sampling schemes. These results are applied in Section 4 to BGW processes with immigration. There we present some open problems too. Finally, in Section 5, we discuss some other directions of statistical inference for branching processes. We have made all efforts to make the exposition self-contained.

2 Classical BGW Branching Processes

Let ξ be a non-negative integer-valued random variable (r.v.) with probability mass function

$$p_k = P(\xi = k), \qquad \sum_{k=0}^{\infty} p_k = 1 \tag{2.1}$$

and probability generating function

$$h(s) = Es^{\xi} = \sum_{k=0}^{\infty} p_k s^k, \quad |s| \leq 1. \tag{2.2}$$

Denote

$$m = E\xi = h'(1) \qquad \text{and} \qquad \sigma^2 = Var\xi = h''(1) - h'(1)(h'(1)-1). \tag{2.3}$$

Let $\{\xi_i(t)\}$ be independent and identically distributed copies of ξ. The classical BGW process is well defined by the recurrence:

$$Z_0 = 1, \qquad Z_{t+1} = \sum_{i=1}^{Z_t} \xi_i(t), \quad t = 0, 1, 2, \ldots \tag{2.4}$$

The r.v. $\xi_i(t)$ can be interpreted as offspring of the ith individual existing in the tth generation. We will call (2.1) – (2.3) individual or offspring characteristics.

Recall the classical trichotomy of the processes: $m < 1$ (*subcritical*), $m = 1$ (*critical*), $m > 1$ (*supercritical*). If $m \leq 1$ then $q = P(Z_t \to 0) = \lim P(Z_t = 0) = 1$, while $m > 1$ implies $q < 1$. The following well known limit results of Kolmogorov (1938) and Yaglom (1947) hold.

Theorem 1. *Let* $\sigma^2 < \infty$ *. Then as* $t \to \infty$*:*
(i) $m < 1 \Rightarrow P(Z_t > 0) \sim Cm^t, Z_t|_{Z_t>0} \xrightarrow{d} Z_\infty$;
(ii) $m = 1 \Rightarrow P(Z_t > 0) \sim 2/(\sigma^2 t), \quad \lim_{t\to\infty} P(2Z_t/\sigma^2 t \le x | Z_t > 0) = 1 - e^{-x}$;
(iii) $m > 1 \Rightarrow W_t = Z_t/m^t \xrightarrow{a.s.} W$ *(also in* L_2*),* $EW = 1$, $q = P(W = 0)$.

Theorem 1 has a lot of refinements and generalizations which can be found in the books cited in the Introduction.

Further on we consider the following **observational schemes** :

- *The entire tree up to time t* :

$$\{\xi_i(k), i = 1, 2, \ldots, Z_k, k = 0, 1, \ldots, t\}. \tag{2.5}$$

- *All successive generations:*

$$(Z_0, Z_1, Z_2, \ldots, Z_t). \tag{2.6}$$

- *Two successive generations:*

$$(Z_t, Z_{t+1}); \tag{2.7}$$

- *The initial and the current generation:*

$$(Z_0, Z_t). \tag{2.8}$$

- *Left censored observations:*

$$(Z_t, Z_{t+1}, \ldots, Z_{t+T}). \tag{2.9}$$

Assume (2.5), that is we have the full information for the process. In this case we can construct the statistic

$$Z_j(k) = \sum_{i=1}^{Z_j} \mathbf{1}(\xi_i(j) = k), \tag{2.10}$$

where $Z_j(k)$ is the number of individuals in the j-th generation having exactly k direct offspring $j = 0, 1, 2, \ldots, t$; $k = 0, 1, 2, \ldots$. Here and further on $\mathbf{1}(\cdot)$ is the indicator function.

Using the independence of the individual evolutions we are able to obtain the likelihood function

$$L_t(p) = \prod_{k=0}^{\infty} p_k^{\sum_{j=0}^{t-1} Z_j(k)},$$

where the parameter $p = (p_0, p_1, p_2, \ldots)$ is the offspring distribution (2.1). It follows from (2.10) that

$$U_t(k) = \sum_{j=0}^{t-1} Z_j(k) \tag{2.11}$$

is the total number of individuals in the first t generations having exactly k offspring and

$$U_t = \sum_{k=0}^{\infty} U_t(k) = \sum_{i=0}^{t-1} Z_i \tag{2.12}$$

is the total number of individuals up to time t (i.e., from 0 to $t-1$ inclusive). Thus, we have to obtain the maximum of the log likelihood

$$\log L_t(p) = \sum_{k=0}^{\infty} U_t(k) \log p_k$$

under the condition

$$\sum_{k=0}^{\infty} p_k = 1, \qquad p_k \geq 0. \tag{2.13}$$

In order to use the Lagrange method we need to introduce the function

$$\Phi_t(p) = \sum_{k=0}^{\infty} U_t(k) \log p_k + \lambda(1 - \sum_{k=0}^{\infty} p_k)$$

and to solve the equations

$$\frac{\partial \Phi_t(p)}{\partial p_k} = U_t(k)/p_k - \lambda = 0, \qquad k = 0, 1, 2, \ldots, \tag{2.14}$$

where λ is the Lagrange multiplier. The solutions of (2.14) are $\widehat{p}_k(t) = U_t(k)/\lambda$ and after plug in them in (2.13) one obtains $\lambda = U_t$. Therefore, the maximum-likelihood estimators (m.l.e.) for $\{p_k\}$ are

$$\widehat{p}_k(t) = U_t(k)/U_t, \qquad k = 0, 1, 2, \ldots \tag{2.15}$$

Note that $\widehat{p}_k(t)$ is the relative proportion of the parents with k offspring. Furthermore, the m.l.e. for m are

$$\widehat{m}_t = \sum_{k=1}^{\infty} k \widehat{p}_k(t). \tag{2.16}$$

Substituting (2.11) and (2.12) in (2.16), it is not difficult to calculate that

$$\widehat{m}_t = (U_{t+1} - Z_0)/U_t = (Z_1 + Z_2 + \cdots + Z_t)/(Z_0 + Z_1 + \cdots + Z_{t-1}) \tag{2.17}$$

Note first that the m.l.e. $\widehat{m}_t$ can be interpreted as a ratio of the "total number of the sons" divided by the "total number of the fathers" (which seems quite natural). On the other hand, surprisingly enough, $\widehat{m}_t$ depends only on the successive generations (2.6). This m.l.e. were proposed by Harris (1948) . It was shown in Dion (1974), Feigin (1977) and Keiding and Lauritzen (1978) that (2.17) is also a m.l.e. for m under the observational scheme (2.6).

Similarly to (2.16), one can consider the m.l.e. for the offspring variance σ^2:

$$\sigma_t^2 = \sum_{k=0}^{\infty} (k - m)^2 \quad \widehat{p}_k(t) \tag{2.18}$$

with known offspring mean m and

$$\widehat{\sigma}_t^2 = \sum_{k=0}^{\infty} (k - \widehat{m}_t)^2 \quad \widehat{p}_k(t) \tag{2.19}$$

with unknown m, respectively. Note that (2.18) and (2.19) cannot be presented in a closed form as (2.17) and depend on the entire branching tree (2.5). It is not difficult to obtain the following useful relation.

$$\sigma_t^2 = \widehat{\sigma}_t^2 + (\widehat{m}_t - m)^2. \tag{2.20}$$

It follows from (2.20) that $\sigma_t^2 \geq \widehat{\sigma}_t^2$ always.

Let us start with the most important estimator $\widehat{m}_t$ for which Harris (1948) proved that it is a consistent estimator for the critical parameter m on the explosion set. Later Heyde (1970) improved this result showing that the Harris estimator is strongly consistent and Dion (1974) obtain that it is also asymptotically normal. Next theorem summarizes these results.

Theorem 2. *Let $m > 1$ and $\sigma^2 < \infty$. Then on the explosion set $\{Z_t \to \infty\}$*

(i) $\widehat{m}_t \to m$ *a.s.;*

(ii) $\left(\sqrt{U_t/\sigma^2}(\widehat{m}_t - m), U_t(m-1)/m^t\right) \xrightarrow{d} (N(0,1), W)$,

where $N(0,1)$ and W are independent r.v. and W is from Theorem 1.

Corollary 1. *Under the conditions of Theorem* 2 *as $t \to \infty$*

(i) $\sqrt{U_t/\sigma^2}(\widehat{m}_t - m) \xrightarrow{d} N(0,1)$;

(ii) $\sqrt{m^t/\sigma^2(m-1)}(\widehat{m}_t - m) \xrightarrow{d} N(0,1)/\sqrt{W}$.

Assuming the observational scheme (2.6), Melnikov and Novikov (1988) introduced the following stopping time estimator:

$$\overline{m}_N = \frac{1}{N}\left(\sum_{k=1}^{\Lambda_N - 1} Z_k + \beta_N\right),$$

where

$$\Lambda_N = \inf\left\{t : \sum_{k=1}^{t} Z_k \geq N\right\}$$

is a stopping time and

$$\beta_N = \left(N - \sum_{k=1}^{\Lambda_N - 1} Z_k\right)/Z_{\Lambda_N - 1}.$$

They obtained the following results.

Theorem 3. *Let $\sigma^2 < \infty$ and $p_0 = 0$. Then $P(\Lambda_N < \infty) = 1$, $E\overline{m}_N = m$ and $Var(\overline{m}_N) \leq \sigma^2/N$.*

In other words, for supercritical processes $(m > 1)$ without extinction the estimator $\overline{m}_N$ is unbiased and consistent as $N \to \infty$.

Assume the observational scheme (2.7). Lotka (1939) introduced the estimator

$$\widetilde{m}_t = (Z_{t+1}/Z_t)\mathbf{1}(Z_t > 0) + \mathbf{1}(Z_t = 0). \tag{2.21}$$

This estimator was first investigated by Nagaev (1967) and also in Bühler (1969), Dion (1974) and Scott (1977). The following results hold.

Theorem 4. *Let $m > 1$ and $\sigma^2 < \infty$. Then on the explosion set $\{Z_t \to \infty\}$*
(i) $E\{\widetilde{m}_t | Z_t > 0\} = m$, $Var\{\widetilde{m}_t | Z_t > 0\} = \sigma^2 E\{Z_t^{-1} | Z_t > 0\}$;
(ii) $\widetilde{m}_t \to m$ *a.s.;*
(iii) $(\sqrt{Z_t/\sigma^2}(\widetilde{m}_t - m), Z_t/m^t) \xrightarrow{d} (N(0,1), W)$,
where $N(0,1)$ and W are independent r.v. and W is from Theorem 1.

Corollary 2. *Under the conditions of Theorem 4, as $t \to \infty$*
(i) $\sqrt{Z_t/\sigma^2}(\widetilde{m}_t - m) \xrightarrow{d} N(0,1)$;
(ii) $\sqrt{m^t/\sigma^2}(\widetilde{m}_t - m) \xrightarrow{d} N(0,1)/\sqrt{W}$.

Under (2.7), Pakes (1975) introduced and studied the following estimators for the offspring probabilities

$$\widetilde{p}_t(k) = (Z_t(k)/Z_t)\mathbf{1}(Z_t > 0) + \mathbf{1}(Z_t = 0), \qquad k = 0, 1, 2, \ldots$$

Suppose now that we can observe only the sample (2.8). In this case Heyde (1975) suggested the moment-type estimator

$$m_t^* = Z_t^{1/t}, \tag{2.22}$$

using the fact that $EZ_t = m^t$.

Theorem 5. *Let $m > 1$ and $\sigma^2 < \infty$. Then on the explosion set $\{Z_t \to \infty\}$*
(i) $m_t^* \to m$ *a.s.;*
(ii) $t(m_t^* - m) \xrightarrow{d} m \log W$.

Note that the estimator m_t^* is consistent but not asymptotically normal.

Another interesting problem is how to estimate the unknown age of the process if we are able to observe only the current size Z_t, where the parameter t is unknown and we have to estimate it. Stigler (1970) proposed the following estimator

$$\widehat{t} = \{\log[Z_t(1-q) - q]\}/\log m. \tag{2.23}$$

Crump and Howe (1972) investigated this problem too.

Theorem 6. *Let $m > 1$. Then on the explosion set $\{Z_t \to \infty\}$ as $t \to \infty$*

$$\widehat{t} - t \to \{\log(1-q)W\}/\log m \qquad a.s.$$

Unfortunately, the estimator (2.23) is not consistent. In the next section we will construct a consistent and asymptotically normal estimator in the noncritical cases and an unbiased estimator in the critical case.

In the case of sample scheme (2.9), Crump and Howe (1972) introduced the following estimator for m

$$\widehat{m}_{t,T} = \sum_{i=t+1}^{t+T} Z_i \Big/ \sum_{i=t}^{t+T-1} Z_i, \tag{2.24}$$

which can be considered as a modification of the Harris estimator ($t = 0$) or the Lotka-Nagaev estimator ($T = 1$). Crump and Howe investigated the cases ($t \to \infty, T$ fixed) and

(t fixed, $T \to \infty$). They proved that $\widehat{m}_{t,T}$ is consistent and asymptotically unbiased estimator in the supercritical case $m > 1$ on the explosion set. In the next section we will consider also the case $t, T \to \infty$ in a more general setting and not only in the supercritical case.

Finally it is interesting to compare the above estimators. It turns out that the best estimator for m is the Harris estimator, which could be expected because it uses more information than the others (see Nitcheva and Yanev (2000) for more details). We will come back to this problem later.

Estimation procedures for σ^2 will be discussed in the following section.

3 BGW Processes with an Increasing Random Number of Ancestors

As it was pointed out in the previous section, the statistical inference for the classical BGW processes is well developed only in the supercritical case and on the explosion set. To avoid that restriction we consider the following generalization of(2.4):

$$Z_{t+1}(n) = \sum_{i=1}^{Z_t(n)} \xi_i(t,n), \qquad t = 0,1,2,\ldots; \quad n = 1,2,\ldots, \tag{3.1}$$

where $\{\xi_i(t,n)\}$ are i.i.d. r.v. with $(2.1)-(2.3)$ and $Z_0(n)$ is an independent of them r.v., which goes to ∞ in some sense. We will refer to the process (3.1) as BGW process with an increasing random number of ancestors (BGWR). Note that the BGWR process satisfies

$$Z_t(n) = \sum_{k=1}^{Z_0(n)} Z_t^{(k)}, t = 0,1,2,\ldots,$$

where $\{Z_t^{(k)}\}$ are i.i.d. copies of the classical BGW process (2.3).

As it was discussed in Yakovlev and Yanev (1989), branching processes with a large and often a random number of ancestors occur naturally in cell proliferation models. Another area of applications are nuclear chain reactions.

In general n and t are independent parameters but very often we will assume that either $n = n(t) \to \infty$ or $t = t(n) \to \infty$. We will construct consistent and asymptotically normal estimators for the offspring probabilities, mean and variance over the entire range of $0 < m < \infty$.

For the estimators related to the process (3.1), we will adopt the notation from Section 2, adding a parameter n, i.e., $\widehat{p}_k(t,n), \widehat{m}_t(n), \sigma_t^2(n), \widehat{\sigma}_t^2(n), \widetilde{m}_t(n), m_t^*(n)$, $\widehat{m}_{t,T}(n)$ etc.

3.1 Estimating the Offspring Mean when $Z_0(n) \equiv n$ a.s.

The case $Z_0(n) \equiv n$ a.s. was first investigated in Yanev (1975) where the asymptotic properties of $\widehat{m}_t(n)$ were studied. We summarize the results in the next four theorems.

Theorem 7. *If $m < \infty$ and $n \to \infty$ then uniformly in $0 \le t \le \infty$ the estimator $\widehat{m}_t(n)$ is strongly consistent and asymptotically unbiased, i.e., $\widehat{m}_t(n) \to m$ a.s. and $E\widehat{m}_t(n) \to m$.*

Denote $c_t = EU_t = \sum_{k=0}^{t-1} m^k$. Also $X_k \sim N(a_k, b_k)$ stands for asymptotic normality, i.e., $(X_k - a_k)/\sqrt{b_k} \xrightarrow{d} N(0,1)$ as $k \to \infty$.

Theorem 8. *Let $\sigma^2 < \infty$ and $t > 0$ be fixed. Then as $n \to \infty$*

$$\widehat{m}_t(n) \sim N(m, \sigma^2/nc_t).$$

If n and t tend to infinity simultaneously then the relation between them is essential for the asymptotic distribution as the following results reveal.

Theorem 9. *Let $\sigma^2 < \infty$ and $n, t \to \infty$.*
(i) If $m < 1$, then $\widehat{m}_t(n) \sim N(m, \sigma^2(1-m)/n)$.
(ii) *If $m = 1$ and $n/t \to \infty$, then $\widehat{m}_t(n) \sim N(m, \sigma^2/nt)$.*
(iii) *If $1 < m < \infty$, then $\widehat{m}_t(n) \sim N(m, \sigma^2(m-1)/nm^t)$.*

Note that by Theorem 9 if $n/t \to \infty$ (as $n, t \to \infty$), then independently of the criticality of the process (i.e., assuming only $m < \infty$) the limiting distribution of $\widehat{m}_t(n)$ is normal (in fact, it is the same as in Theorem 8).

In the critical case the estimator has a more complex asymptotic behavior as the following result shows.

Theorem 10. *Let $m = 1$, $\sigma^2 < \infty$, and $n^2/t \to 0$ as $n, t \to \infty$. Then*

$$(2n/\sigma^2)(1 - \widehat{m}_t(n)) \xrightarrow{d} 1/Y,$$

where the r.v. Y has a stable distribution with parameter $1/2$, i.e., its Laplace transform is $Ee^{-\lambda Y} = \exp\{-\lambda^{1/2}\}$.

Note that the critical case when $O(n) \le t \le O(n^2)$ as $n \to \infty$, is still an open problem. Yanev (1986) obtains further results using random normalizing constants.

3.2 Limiting Distributions

In a series of papers Dion and Yanev (1991-1997) study the general case of random $Z_0(n)$. First we will show that the limiting behavior of the BGWR process (3.1) differs significantly from that of the classical BGW process (2.3) as given in Theorem 1. From now on we will suppose that ν is a positive r.v. with d.f. $F_\nu(x)$.

Theorem 11. *Let $m < 1$, $\sigma^2 < \infty$, $Z_0(n)/n \xrightarrow{d} \nu$, and $n, t \to \infty$.*
(i) *If $nm^t \to 0$, then $Z_t(n) \xrightarrow{P} 0$.*
(ii) *If $nm^t \to C_1$, $0 < C_1 < \infty$, then*

$$Z_t(n) \xrightarrow{d} \eta_1,$$

where $Ee^{-\lambda\eta_1} = E\exp\{-\alpha_1\nu(1-g(e^{-\lambda})\}$, $\lambda > 0$ and the p.g.f. $g(s)$ is the unique solution of the equation $g(h(s)) = 1 - m(1 - g(s))$ and $\alpha_1 = C_1/g'(1)$.

Theorem 12. *Let* $m = 1$, $\sigma^2 < \infty$, $Z_0(n)/n \xrightarrow{d} \nu$, *and* $n,t \to \infty$.

(i) *If* $n/t \to 0$, *then* $Z_t(n) \xrightarrow{P} 0$.

(ii) *If* $n/t \to C_2$, $0 < C_2 < \infty$, *then*

$$Z_t(n)/n \xrightarrow{d} \eta_2,$$

where $Ee^{-\lambda\eta_2} = E\exp\{-\lambda\alpha_2\nu/(\alpha_2+\lambda)\}$, $\lambda > 0$, *and* $\alpha_2 = 2C_2/\sigma^2$.

Note that if $\nu = 1$ a.s. then η_2 has a compound Poisson distribution.

Further on we make the following assumption

$$(m > 1) \vee (m = 1, n/t \to \infty) \vee (m < 1, nm^t \to \infty), \tag{3.2}$$

which covers all cases not included in Theorem 11 and Theorem 12.

Theorem 13. *Assume (3.2),* $\sigma^2 < \infty$, $Z_0(n)/n \xrightarrow{d} \nu$, *and* $n,t \to \infty$.

(i) *For every* $n \geq 1$, $W(n) = \{W_t(n) = Z_t(n)/(Z_0(n)m^t), t \geq 1\}$ *is a martingale and* $W_t(n) \xrightarrow{P} 1$ *(Law of Large Numbers).*

(ii) $\sqrt{Z_0(n)A_t}(W_t(n) - 1) \xrightarrow{d} N(0,1)$ *(Central Limit Theorem),*

where

$$A_t = m^t \mathbf{1}(m < 1) + \frac{1}{\sigma^2 t}\mathbf{1}(m = 1) + \frac{m(m-1)}{\sigma^2}\mathbf{1}(m > 1). \tag{3.3}$$

3.3 Harris Estimators

Let us recall the definitions (2.15)-(2.19) from Section 2 and apply them to the BGWR process.

Theorem 14. *Let* $n,t \to \infty$.

(i) *If* $m < \infty$ *and* $Z_0(n) \xrightarrow{P} \infty$, *then* $\widehat{m}_t(n) \xrightarrow{P} m$ *and* $\widehat{p}_k(t,n) \xrightarrow{P} p_k$, $k = 0,1,\ldots$

(ii) *If* $\sigma^2 < \infty$ *and* $Z_0(n)/n \xrightarrow{d} \nu$, *then* $E\widehat{m}_t(n) \to m$.

Note that if $Z_0(n) \to \infty$ a.s., then $\widehat{m}_t(n) \to m$ a.s. and $\widehat{p}_k(t,n) \to p_k$ a.s. for $k \geq 0$.

Let $\widehat{\theta}_t(n)$ be one of the following statistics:

$$(\widehat{p}_k(t,n) - p_k)/\sqrt{p_k(1-p_k)},$$

$$(\widehat{m}_t(n) - m)/\sigma,$$

$$(\sigma_t^2(n) - \sigma^2)/D,$$

$$(\widehat{\sigma}_t^2(n) - \sigma^2)/D,$$

where $D = Var(\xi - m)^2 = E(\xi - m)^4 - \sigma^4$.

Theorem 15. *Let* $\sigma^2 < \infty$ *and* $Z_0(n)/n \xrightarrow{P} \nu$. *If* $n,t \to \infty$ *(and additionally* $n/t \to \infty$ *when* $m = 1$*), then*

$$(\sqrt{U_t(n)}\widehat{\theta}_t(n), U_t(n)/nc_t) \xrightarrow{d} (N(0,1), \widetilde{\nu}),$$

where $N(0,1)$ *and* $\widetilde{\nu}$ *are independent random variables and* $\widetilde{\nu}$ *is identically distributed as* ν. *Of course, in the case when* $\theta_t(n)$ *is an estimator for the variance* σ^2 *we need the additional assumption* $E\xi^4 < \infty$.

Corollary 3. *Under the conditions of Theorem* 15

(i) $\sqrt{U_t(n)}\widehat{\theta}_t(n) \xrightarrow{d} N(0,1)$.

(ii) $\sqrt{nc_t}\widehat{\theta}_t(n) \xrightarrow{d} N(0,1)\sqrt{\nu}$.

Duby and Rouault (1980,1982) studied the joint asymptotic distribution of $(\widehat{m}_t(n), (\widehat{\sigma}_t^2(n))$ when $m > 1$ and $Z_0(n) \equiv n$ a.s. They obtained asymptotic normality for two related independent random variables with different normalizing factors.

The asymptotic behavior in the critical case is more complex. The following generalization of Theorem 10 holds.

Theorem 16. *Let* $m = 1$, $\sigma^2 < \infty$, $Z_0(n)/n \xrightarrow{d} \nu$, *and* $n^2/t \to 0$ *as* $n, t \to \infty$. *Then*

(i) $\sqrt{2U_t(n)/\sigma^2}\,(1 - \widehat{m}_t(n)) \xrightarrow{d} |N(0,1)|$.

(ii) $(2n/\sigma^2)(1 - \widehat{m}_t(n)) \xrightarrow{d} 1/X$, *where* $Ee^{-\lambda X} = E\exp\{-\sqrt{\lambda\nu}\}$.

If $\nu = 1$ a.s. then X has a stable distribution with parameter $1/2$. Notice also that the case $O(n) \leq t \leq O(n^2), n \to \infty$ when $m = 1$ is still an open problem.

The statement in part (i) shows the asymptotic normality of $\widehat{\theta}_t(n)$ (under random normalization) when $n/t \to \infty$ (as $n, t \to \infty$) uniformly in $0 < m < \infty$. This fact allows us to construct asymptotic confidence intervals (a.c.i.) and to test hypotheses for the process' criticality.

Indeed, let us pick a confidence level $1-\alpha$ and let $\Phi(\gamma_{1-\alpha/2}) = 1 - \alpha/2$. Then the a.c.i. for m and p_k $(k = 0, 1, 2, \ldots)$ are given as follows:

(i) $(\widehat{m}_t(n) \pm \sigma\gamma_{1-\alpha/2}/\sqrt{U_t(n)})$;

(ii) $(\{\widehat{p}_k(t,n) + \gamma_{1-\alpha/2}^2\}/\{1 + \gamma_{1-\alpha/2}^2/U_t(n)\}$
$\pm\sqrt{\gamma_{1-\alpha/2}^4 + 4U_t(n)\gamma_{1-\alpha/2}^2\widehat{p}_k(t,n)(1 - \widehat{p}_k(t,n))}/2\{U_t(n) + \gamma_{1-\alpha/2}^2\})$.

Similarly, applying Theorem 16, one can obtain a.c.i. when $m = 1$ and $n^2/t \to 0$.

3.4 Lotka-Nagaev Estimators

Assume the sampling scheme (2.7) and let $\widetilde{\theta}_t(n) = (\widetilde{m}_t(n) - m)/\sigma$ where $\widetilde{m}_t(n)$ is defined in (2.21) for a BGWR process.

Theorem 17. *Assume* $\sigma^2 < \infty$, *condition* (3.2), *and* $n, t \to \infty$.

(i) *If* $Z_0(n)/n \xrightarrow{P} \nu$, *then*

$$(\sqrt{Z_t(n)}\,\widetilde{\theta}_t(n), Z_t(n)/nm^t) \xrightarrow{d} (N(0,1), \widetilde{\nu})$$

and

$$\sqrt{nm^t}\widetilde{\theta}_t(n) \xrightarrow{d} N(0,1)/\sqrt{\widetilde{\nu}},$$

where $N(0,1)$ *and* $\widetilde{\nu}$ *are independent r.v. and* $\widetilde{\nu}$ *is identically distributed as* ν.

(ii) *If* $Z_0(n)/n \xrightarrow{d} \nu$, *then*

$$\sqrt{Z_t(n)}\,\widetilde{\theta}_t(n) \xrightarrow{d} N(0,1)$$

and

$$\sqrt{Z_0(n)m^t}\,\widetilde{\theta}_t(n) \xrightarrow{d} N(0,1).$$

Let us compare Theorem 17 with Theorem 4. First of all, Theorem 4 deals with the supercritical case only. Moreover, in Theorem 4 with the non-random normalizing factors, the limit includes the nuisance random variable W.

3.5 Heyde Moment-Type Estimators

Assume the sampling scheme (2.8) and let

$$m_t^*(n) = [Z_t(n)/Z_0(n)]^{1/t}.$$

Theorem 18. *Suppose that* $Z_0(n)/n \xrightarrow{d} \nu$, $\sigma^2 < \infty$, *and condition* (3.2) *holds with* $n,t \to \infty$. *Then*

(i) $m_t^*(n) \xrightarrow{P} m.$

(ii) $\frac{t}{n}\sqrt{Z_0(n)A_t}(m_t^*(n) - m) \xrightarrow{d} N(0,1)$ *where* A_t *is defined in* (3.3).

Note that Theorem 18 establishes consistency and asymptotic normality not only in the supercritical case but also in the critical and subcritical cases, while Theorem 5 from Section 2 applies to the supercritical case only and the limiting distribution there is not normal.

3.6 Estimating the Unknown Age of the Process

The problem is how to determine the age of the process when only its size is observed. In other words, we observe the population size $Z_t(n)$ but the population index t is unknown and we would like to estimate it. Define the estimators:

$$\widehat{t}(n) = \log(Z_t(n)/Z_0(n))/\log m \qquad \text{if} \qquad m \neq 1; \tag{3.4}$$

$$\widehat{T}(n) = (Z_t(n) - Z_0(n))^2/Z_0(n)\sigma^2 \qquad \text{if} \qquad m = 1. \tag{3.5}$$

Theorem 19. *Let* $\sigma^2 < \infty$, $Z_0(n)/n \xrightarrow{d} \nu$, *and* $n,t \to \infty$.

(i) *If* $(m > 1) \vee (m < 1, nm^t \to \infty)$, *then* $\widehat{t}(n) - t \xrightarrow{P} 0$.

(ii) *If* $m > 1$ *then*

$$(\log m)\sqrt{Z_0(n)m(m-1)/\,\sigma^2}(\widehat{t}(n) - t) \xrightarrow{d} N(0,1);$$

(iii) *If* $m < 1$ *and* $nm^t \to \infty$, *then*

$$(\log m)\sqrt{Z_0(n)m^t}(\widehat{t}(n) - t) \xrightarrow{d} N(0,1);$$

(iv) *If* $m = 1$ *then* $E\widehat{T}(n) = t$; *if additionally* $n/t \to \infty$ *then*

$$\widehat{T}(n)/t \xrightarrow{d} \chi_1^2 = [N(0,1)]^2.$$

Theorem 19 is a counterpart of Theorem 6 from Section 2.

3.7 Conditional Least Squares Estimators

From (3.1) it is not difficult to obtain that for every $k \geq 1$

$$E\{Z_k(n)|Z_{k-1}(n)\} = mZ_{k-1}(n) \tag{3.6}$$

and

$$E\{[Z_k(n) - mZ_{k-1}(n)]^2|Z_{k-1}(n)\} = \sigma^2 Z_{k-1}(n). \tag{3.7}$$

From (3.6) it follows that one can consider

$$Z_k(n) = mZ_{k-1}(n) + \alpha_k(n) \tag{3.8}$$

as a stochastic regression equation with the unknown coefficient m and the martingale difference error term $\alpha_k(n)$. In view of (3.7) it is most convenient to present $\alpha_k(n)$ as $\varepsilon_k(n)\sqrt{Z_{k-1}(n)}$. Now minimizing $S^2_{t,n}(m) = \sum_{k=1}^t \varepsilon_k^2(n)$ with respect to m, one obtains the least squares estimator (l.s.e.) for m

$$\widehat{m}_t(n) = \sum_{k=1}^t Z_k(n) / \sum_{k=1}^t Z_{k-1}(n), \tag{3.9}$$

which is simply the m.l. Harris estimator, studied in Section 3.3.

Similarly, by (3.7), one can consider the stochastic regression equation

$$[Z_k(n) - mZ_{k-1}(n)]^2 = \sigma^2 Z_{k-1}(n) + \beta_k(n) \tag{3.10}$$

with parameters m and σ^2 and the martingale difference error term $\beta_k(n)$.

Let

$$Y_k(n;x) = Z_{k-1}(n)[Z_k(n)/Z_{k-1}(n) - x]^2 \mathbf{1}(Z_{k-1}(n) > 0). \tag{3.11}$$

Let us set $\beta_k(n) = \varepsilon_k(n) Z_{k-1}(n)$. Then by (3.10) and (3.11) we have

$$Y_k(n;m) = \sigma^2 + \varepsilon_k(n).$$

To obtain the l.s.e. for σ^2 we need to minimize $S^2_{t,n}(\sigma^2) = \sum_{k=1}^t \varepsilon_k^2(n)$ with respect to σ^2. When m is known we have

$$\sigma_t^2(n;m) = \frac{1}{t} \sum_{k=1}^t Y_k(n;m), \tag{3.12}$$

and similarly, when m is unknown,

$$\widehat{\sigma}_t^2(n;\widehat{m}) = \frac{1}{t} \sum_{k=1}^t Y_k(n;\widehat{m}_t(n)). \tag{3.13}$$

The following theorem shows that the estimators (3.12) and (3.13) are consistent and asymptotically normal.

Theorem 20. *Assume* $E\xi^4 < \infty$ *and condition* (3.2).

(i) *If* $Z_0(n)/n \xrightarrow{d} \nu$, *then*

$$\sigma_t^2(n;m) \xrightarrow{P} \sigma^2 \quad \text{and} \quad \sigma_t^2(n;m) \sim N(\sigma^2, 2\sigma^4/t)$$

(ii) *If* $Z_0(n)/n \xrightarrow{P} \nu$, *then*

$$\widehat{\sigma}_t^2(n;\widehat{m}) \xrightarrow{P} \sigma^2 \quad \text{and} \quad \widehat{\sigma}_t^2(n;\widehat{m}) \sim N(\sigma^2, 2\sigma^4/t).$$

Heyde (1974) and Dion (1975) investigate the case $Z_0(n) \equiv 1$ a.s. and $m > 1$. Dion proves that on the explosion set $\sigma_t^2(\widehat{m}_t)$ is consistent and asymptotically normal estimator for σ^2. Heyde obtains similar results for $\sigma_t^2(\widetilde{m}_t)$, where $\widetilde{m}_t$ is the Lotka-Nagaev estimator (2.21).

3.8 Censored Estimators

Consider $(Z_t(n), Z_{t+1}(n), \ldots, Z_{t+T}(n))$ for some t and T, which can be interpreted as a left censored sample. For m we will use the estimator (2.24) as it applies to (3.1), i.e.,

$$\widehat{m}_{t,T}(n) = \sum_{i=t+1}^{t+T} Z_i(n) / \sum_{i=t}^{t+T-1} Z_i(n). \tag{3.14}$$

For σ^2 we construct the estimators

$$\sigma_{t,T}^2(n;m) = \frac{1}{T} \sum_{k=t+1}^{t+T} Y_k(n;m), \tag{3.15}$$

if m is known and

$$\widehat{\sigma}_{t,T}^2(n;\widehat{m}) = \frac{1}{T} \sum_{k=t+1}^{t+T} Y_k(n;\widehat{m}_t(n)), \tag{3.16}$$

if m is unknown.

Note that for $t = 0$ the estimator (3.15) is equivalent to (3.12) and (3.16) to (3.13).

Theorem 21. *Assume* (3.2), $\sigma^2 < \infty$, *and* $Z_0(n)/n \xrightarrow{d} \nu$. *If* $n, t \to \infty$ *then, uniformly in* $1 \leq T \leq \infty$, *we have*

(i) $\widehat{m}_{t,T}(n) \xrightarrow{P} m$ *and* $E\,\widehat{m}_{t,T}(n) \to m$.

(ii) $\sqrt{Z_t(n) + \cdots + Z_{t+T-1}(n)}(\widehat{m}_{t,T}(n) - m)/\sigma \xrightarrow{d} N(0,1)$.

Theorem 22. *Assume* (3.2), $E\xi^4 < \infty$ *and* $Z_0(n)/n \xrightarrow{P} \nu$. *If* $n, T \to \infty$ *then, uniformly in* $1 \leq t \leq \infty$ *we have*

(i) $\sigma_{t,T}^2(n;m) \xrightarrow{P} \sigma^2$, $E\sigma_{t,T}^2(n;m) \to \sigma^2$, $\sqrt{2T/\sigma^2}(\sigma_{t,T}^2(n;m) - \sigma^2) \xrightarrow{d} N(0,1)$.

(ii) $\widehat{\sigma}_{t,T}^2(n;\widehat{m}) \xrightarrow{P} \sigma^2$, $E\widehat{\sigma}_{t,T}^2(n;\widehat{m}) \to \sigma^2$, $\sqrt{2T/\sigma^2}(\widehat{\sigma}_{t,T}^2(n;\widehat{m}) - \sigma^2) \xrightarrow{d} N(0,1)$.

Finally, as it was pointed out at the end of Section 2, we are in a position to complete the results of Crump and Howe (1972) . Namely, let us consider the case $Z_0(n) \equiv 1$ a.s. In this case (3.14) and (2.24) are identical. Similarly, for (3.15) and (3.16) we use $\sigma_{t,T}^2(m)$ and $\widehat{\sigma}_{t,T}^2(\widehat{m})$.

Theorem 23. *Let* $m > 1$ *and* $\sigma^2 < \infty$. *Then, under the condition of non-extinction, we have*

(i) *If* $t \to \infty$ *then, uniformly in* $1 \leq T \leq \infty$, *the estimator* $\widehat{m}_{t,T}$ *is consistent, asymptotically unbiased and asymptotically normal, i.e.,*

$$\widehat{m}_{t,T} \xrightarrow{P} m, \quad E\widehat{m}_{t,T} \to m, \quad \sqrt{Z_t + \cdots + Z_{t+T-1}}(\widehat{m}_{t,T} - m)/\sigma \xrightarrow{d} N(0,1).$$

(ii) *If* $E\xi^4 < \infty$ *and* $t, T \to \infty$, *then* $\sigma_{t,T}^2(m)$ *and* $\sigma_{t,T}^2(\widehat{m})$ *are consistent and asymptotically normal estimators for* σ^2, *i.e.,*

$$\sigma_{t,T}^2(m) \xrightarrow{P} \sigma^2 \widehat{\sigma}_{t,T}^2(\widehat{m}) \xrightarrow{P} \sigma^2, \quad \sigma_{t,T}^2(m) \sim N(0, 2\sigma^4/T), \quad \widehat{\sigma}_{t,T}^2(\widehat{m}) \sim N(0, 2\sigma^4/T).$$

The proofs of Theorems 7–8 and Theorems 14–17 are given in Dion and Yanev (1997). They use random summation methods and a generalization of the Gnedenko-Fahim's transfer limit theorem. Theorems 9, 18 and 19 are proved in Dion and Yanev (1995) and Theorems 20-23 - in Dion and Yanev (1994).

4 BGW Processes with Immigration

Traditionally, the branching processes with or without immigration have been treated separately. In this section we will show that the estimation theory developed in the previous section can be transferred to processes with immigration designing a suitable sampling scheme.

It is well known that the BGW process with immigration can be defined as follows

$$Y_0 = 0, \quad Y_{t+1} = \sum_{i=1}^{Y_t} \xi_i(t) + I_{t+1}, \quad t = 0, 1, 2, \ldots, \tag{4.1}$$

where the offspring $\{\xi_i(t)\}$ and the immigration component $\{I_t\}$ are independent. We assume that r.v. $\{\xi_i(t)\}$ are i.i.d. and have the same individual characteristics as in Section 2 (2.1)-(2.3). Also, let $\{I_t\}$ be non-negative integer valued i.i.d. r.v. with

$$\lambda = EI_t, \qquad b^2 = VarI_t. \tag{4.2}$$

The following limit results are conterparts of those in Theorem 1.

Theorem 24. *Let $\sigma^2 < \infty$ and $t \to \infty$. Then*

(i) $m < 1 \Rightarrow Y_t \xrightarrow{d} Y_\infty, \qquad EY_\infty = \lambda/(1-m)$.

(ii) $m = 1 \Rightarrow \lim P(Y_t/\beta t \le x) = \frac{1}{\Gamma(\alpha)} \int\limits_0^x u^{\alpha-1} e^{-u} du$, *where $\beta = \sigma^2/2$ and $\alpha = \lambda/\beta$. If $\lambda \le \beta$, then the Markov chain $\{Y_t\}$ is null recurrent; it is transient if $\lambda > \beta$.*

(iii) $m > 1 \Rightarrow V_t = Y_t/m^t \to^{a.s.} V$ *(and also in L_2)*, $\quad EV = \lambda m/(m-1)$.

The first branching process with immigration was investigated by Sevastyanov (1957) in the continuous time Markov case. The discrete time model (4.1) was introduced by Heathcote (1965, 1966). Theorem 24 has a lot of refinements and generalizations some of which can be found in the books cited in the Introduction.

Similarly as in Section 3.7, it is not difficult to obtain from (4.1) that for every $k \ge 1$

$$E\{Y_k | Y_{k-1}\} = mY_{k-1} + \lambda. \tag{4.3}$$

From (4.3) it follows that

$$Y_k = mY_{k-1} + \lambda + \alpha_k \tag{4.4}$$

is stochastic regression equation with unknown parameters m and λ and a martingale difference error term α_k. Minimizing $S_t^2(m,\lambda) = \sum_{k=1}^t \alpha_k^2$ with respect to m and λ, we obtain the conditional l.s.e.

$$\widehat{m}_t = \left\{ \sum_{k=1}^t Y_k \sum_{k=0}^{t-1} Y_k - t \sum_{k=1}^t Y_k Y_{k-1} \right\} \Big/ \left\{ \left(\sum_{k=0}^{t-1} Y_k \right)^2 - t \sum_{k=0}^{t-1} Y_k^2 \right\}, \tag{4.5}$$

$$\widehat{\lambda}_t = \left\{ \sum_{k=0}^{t-1} Y_k \sum_{k=1}^{t} Y_k Y_{k-1} - \sum_{k=1}^{t} Y_k \sum_{k=0}^{t-1} Y_k^2 \right\} \Big/ \left\{ \left(\sum_{k=0}^{t-1} Y_k \right)^2 - t \sum_{k=0}^{t-1} Y_k^2 \right\}. \quad (4.6)$$

The estimators (4.5) and (4.6) were introduced and investigated in the subcritical case by Heyde and Seneta (1972, 1974). The subcritical case was also studied by Quine (1976, 1977).

Theorem 25. *Let $\sigma^2 < \infty$ and $b^2 < \infty$. If $m < 1$ then $\widehat{m}_t$ and $\widehat{\lambda}_t$ are strongly consistent and asymptotically normal estimators for m and λ.*

The estimation of the variances σ^2 and b^2 was first discussed and investigated by Yanev and Tchoukova-Dantcheva (1980, 1986). Since

$$E\{U_k^2 | Y_{k-1}\} = \sigma^2 Y_{k-1} + b^2, \quad (4.7)$$

where $U_k = Y_k - mY_{k-1} + \lambda$, we have

$$U_k^2 = \sigma^2 Y_{k-1} + b^2 + \beta_k \quad (4.8)$$

i.e., a stochastic regression equation with unknown parameters m, λ, σ^2 and b^2, and a martingale difference error term β_k.

Assuming that m and λ are known and minimizing $S_t^2(\sigma^2, b^2) = \sum_{k=1}^{t} \beta_k^2$ with respect to σ^2 and b^2, we obtain the conditional l.s.e.

$$\overline{\sigma}_t^2 = \sigma_t^2(m, \lambda) = \sum_{k=1}^{t} U_k^2 (Y_k - \overline{Y}_t) \Big/ \sum_{k=1}^{t} (Y_k - \overline{Y}_t)^2, \quad (4.9)$$

$$\overline{b}_t^2 = b_t^2(m, \lambda) = \sum_{k=1}^{t} U_k^2 (Y_k^2 - Y_k \overline{Y}_t) \Big/ t \sum_{k=1}^{t} (Y_k - \overline{Y}_t)^2, \quad (4.10)$$

where $\overline{Y}_t = \frac{1}{t} \sum_{k=1}^{t} Y_k$.

If the means m and λ are unknown we consider the estimators

$$\widehat{\sigma}_t^2 = \sigma_t^2(\widehat{m}_t, \widehat{\lambda}_t) \quad \text{and} \quad \widehat{b}_t^2 = b_t^2(\widehat{m}_t, \widehat{\lambda}_t), \quad (4.11)$$

which are obtained by (4.9) and (4.10), replacing U_k with $\widehat{U}_k = Y_k - \widehat{m}_t Y_{k-1} + \widehat{\lambda}_t$.

Theorem 26. *Let $E\xi^4 < \infty$ and $EI_t^4 < \infty$. If $m < 1$ then the estimators* (4.9)–(4.11) *are consistent and asymptotically normal for σ^2 and b^2.*

Setting $\alpha_k = \varepsilon_k / \sqrt{Y_k + 1}$ in (4.4) and minimizing $\sum_{k=1}^{t} \varepsilon_k^2$ we obtain the following weighted conditional l.s.e. for m and λ

$$\widetilde{m}_t = \left\{ \sum_{k=1}^{t} Y_k \sum_{k=0}^{t-1} X_k - t \sum_{k=1}^{t} Y_k X_{k-1} \right\} \Big/ \left\{ \sum_{k=0}^{t-1} (1 + Y_k) \sum_{k=0}^{t-1} X_k - t^2 \right\},$$

$$\widetilde{\lambda}_t = \left\{ \sum_{k=0}^{t-1} Y_k \sum_{k=1}^{t} Y_k X_{k-1} - \sum_{k=1}^{t} Y_k \sum_{k=0}^{t-1} Y_k X_k \right\} \Big/ \left\{ \sum_{k=0}^{t-1} (1 + Y_k) \sum_{k=0}^{t-1} X_k - t^2 \right\},$$

where $X_k = 1/(Y_k + 1)$.

Similarly, by (4.8,) the weighted conditional l.s.e. for σ^2 and b^2 are

$$\widetilde{\sigma}_t^2 = \left\{ \sum_{k=1}^{t} \widetilde{U}_k^2 X_{k-1} \sum_{k=0}^{t-1} X_k^2 - \sum_{k=1}^{t} \widetilde{U}_k^2 X_{k-1}^2 \sum_{k=0}^{t-1} X_k \right\} \Big/ \left\{ t \sum_{k=0}^{t-1} X_k^2 - \left(\sum_{k=0}^{t-1} X_k \right)^2 \right\},$$

$$\widetilde{b}_t^2 = \left\{ \sum_{k=1}^{t} \widetilde{U}_k^2 X_{k-1}^2 \sum_{k=0}^{t-1} Y_k X_k - \sum_{k=1}^{t} \widetilde{U}_k^2 X_{k-1} \sum_{k=0}^{t-1} Y_k X_k^2 \right\} \Big/ \left\{ t \sum_{k=0}^{t-1} X_k^2 - \left(\sum_{k=0}^{t-1} X_k \right)^2 \right\},$$

where $\widetilde{U}_k = Y_k - \widetilde{m}_k Y_{k-1} + \widetilde{\lambda}_k$.

The weighted conditional l.s.e. were introduced and studied by Wei and Winnicki (1989, 1990) and Winnicki (1988, 1991).

Theorem 27. *Let $\sigma^2 < \infty$, $b^2 < \infty$, and $t \to \infty$.*

(i) *If $m \neq 1$ then $\widetilde{m}_t \to^{a.s.} m$; if $m = 1$ then $\widetilde{m}_t \xrightarrow{P} m$.*

(ii) *If $m < 1$ then $\widetilde{\lambda}_t \to^{a.s.} \lambda$; if $m = 1$ then $\widetilde{\lambda}_t \xrightarrow{P} \lambda$: if $m > 1$ then there is no consistent estimator for λ;*

(iii) *If $m > 1$ then $\widetilde{\theta}_t = \sqrt{t(\overline{Y}_t + 1)}(\widetilde{m}_t - m) \xrightarrow{d} N(0, \sigma^2)$;*

(iv) *If $m < 1$ then $\widetilde{\theta}_t \xrightarrow{d} N(0, C^2)$, where the constant C^2 is explicitly calculated;*

(v) *If $m = 1$ then $\widetilde{\theta}_t \xrightarrow{d} (\zeta(1) - \lambda)(\int_0^1 \zeta(t)dt)^{-1/2}$, where $\zeta(t)$ is a diffusion process with generator $Af(x) = \beta x f''(x) + \lambda f'(x)$.*

Theorem 28. *Let $E\xi^4 < \infty$, $EI_t^4 < \infty$, and $t \to \infty$.*

(i) *If $m \neq 1$ then $\widetilde{\sigma}_t^2 \to^{a.s.} \sigma^2$; If $m = 1$ then $\widetilde{\sigma}_t^2 \xrightarrow{P} \sigma^2$;*

(ii) *If $m < 1$ then $\widetilde{b}_t^2 \to^{a.s.} b^2$;*

(iii) *If $m = 1$ and $\lambda \leq \beta$ then $\widetilde{b}_t^2 \xrightarrow{P} b^2$;*

(iv) *If $m = 1$ and $\lambda > \beta$; or $m > 1$ then the parameter b^2 can not be estimated.*

Now we will discuss the papers of Dion and Yanev (1995, 1997) where it is shown how to transfer the asymptotic results of the BGW process with a random number of ancestors (BGWR) to the BGW process with immigration (BGWI).

Consider the branching tree underlining the process $\{Y_t\}$ and let $Z_t(n)$ be the number of individuals, among generations $t+1, \ldots, t+n$, whose ancestors immigrated exactly t generations ago. This type of data may be available in demography or cell biology, for example. Observe that $Z_0(n) = \sum_{k=1}^{n} I_k$ is the total number of immigrants in the first n generations, $Z_1(n)$ is the total number of their offsprings and so on. Therefore $\{Z_t(n)\}$ is a BGWR process investigated in Section 3. In fact, $Z_t(n)$ is the cardinality of a particular "diagonal stopping line" over the sampling tree of the BGWI process, while Y_t is the cardinality of a "horizontal stopping line" (see Chauvin (1986) for the notion of stopping line). This duality is very important and can be exploited to transfer some results and properties between the two types branching processes. Namely, to estimate the offspring characteristics in a branching process with immigration one has to consider the proposed "diagonal" sampling scheme i.e., the observations $(Z_0(n), Z_1(n), Z_2(n), \ldots, Z_t(n))$ (or even part of them). Then applying the results of Section 3 we can estimate m and σ^2.

Assume that it is possible to observe $Z_j(n,k)$ too, i.e., the number of individuals in the first $(j+n)$ generations of the process (4.1) having exactly k offspring in the next generation and whose ancestors immigrated exactly j generations ago. Then $U_t(n,k) = \sum_{j=0}^{t-1} Z_j(n,k)$ is the total number of the individuals in the first $(n+t)$ generations of the process (4.1) having k offspring in the next generation and whose ancestors immigrated up to t generations ago. Now we can calculate $U_t(n) = \sum_{k=0}^{\infty} U_t(n,k) = \sum_{i=0}^{t-1} Z_i(n)$, which is the total number of the individuals of the first $(n+t)$ generations of process (4.1) which ancestors immigrated up to t generations ago. Therefore one can obtain the estimator $p_k(t,n) = U_t(n,k)/U_t(n)$ of the offspring probability $p_k, k = 0,1,2,\ldots,$ and one can apply Theorems 14 and 15.

Note that in the case (4.1) the r.v. $\{I_t\}$ are i.i.d. Therefore by the LLN $Z_0(n)/n = \frac{1}{n}\sum_{k=1}^{n} I_k \to^{a.s.} \lambda$. This way applying the results of Section 3 we obtain strongly consistent and asymptotically normal estimators for the offspring characteristics.

Notice also that for the estimation theory of Section 3 we need only that $Z_0(n)/n = \frac{1}{n}\sum_{k=1}^{n} I_k \to \nu$ (in probability on even in distribution for some cases), where ν is a positive random variable. This is a significant extension of the model (4.1). Applying the results from Section 3 we are able to estimate m and σ^2 even if some assumptions of Theorems 24-28 do not hold.

Suppose that in (4.1) $\{I_t\}$ are i.i.d. In this case the following estimators for λ are proposed

$$\widetilde{\lambda}_t(n) = Z_t(n)/nm^t \quad \text{and} \quad \widehat{\lambda}_t(n) = Z_t(n)/n[\widehat{m}_t(n)]^t, \tag{4.12}$$

where $\widehat{m}_t(n)$ is the Harris estimator from Section 3.

Theorem 29. *Assume $\sigma^2 < \infty$ and condition (3.2). Then as $n,t \to \infty$*

$$\widetilde{\lambda}_t(n) \xrightarrow{P} \lambda \quad \textit{and} \quad \widehat{\lambda}_t(n) \xrightarrow{P} \lambda.$$

The theorem is still valid if $\{I_t\}$ is a stationary ergodic process for which $\frac{1}{n}\sum_{k=1}^{n} I_k$ converges in probability to some positive constant λ.

Theorem 30. *Assume $\sigma^2 < \infty$, $b^2 < \infty$ and $n,t \to \infty$.*

(i) *If $m < 1$ and $nm^t \to \infty$ then*

$$\sqrt{nm^t}(\widetilde{\lambda}_t(n) - \lambda) \xrightarrow{d} N(0,\lambda) \quad \textit{and} \quad \sqrt{nm^t}(\widehat{\lambda}_t(n) - \lambda) \xrightarrow{d} N(0,\lambda);$$

(ii) *If $m = 1$ and $n/t \to \infty$ then*

$$\sqrt{n/t}(\widetilde{\lambda}_t(n) - \lambda) \xrightarrow{d} N(0,\lambda\sigma^2) \quad \textit{and} \quad \sqrt{n/t}(\widehat{\lambda}_t(n) - \lambda) \xrightarrow{d} N(0,b^2);$$

(iii) *If $m > 1$ then*

$$\sqrt{n}(\widetilde{\lambda}_t(n) - \lambda) \xrightarrow{d} N(0,\lambda\sigma^2/m(m-1) + b^2)$$

and

$$\sqrt{n}(\widehat{\lambda}_t(n) - \lambda) \xrightarrow{d} N(0,\lambda\sigma^2/m(m-1) + b^2).$$

Atanasov et al. (2007) study a robust modifications of the estimators (4.12) and also make a comparisons with $\widehat{\lambda}_t$ and $\widetilde{\lambda}_t$.

Note that Wei and Winnicki (1990) show that, in the supercritical case, the immigration mean λ does not admit a consistent estimator on the basis of $\{Y_t\}$ alone (see Theorem 27(ii)).

In general, the knowledge of $\{(Y_0, Y_{1,}, \ldots, Y_{t+n}), \sum_{k=1}^{n} I_k\}$ seems to be asymptotically equivalent to $\{Z_0(n), Z_1(n), \ldots, Z_t(n)\}\}$ as $n, t \to \infty$, at least on the set of non-extinction. Let us point out that this idea is not exploited.

5 Concluding Remarks and Comments

In this chapter we have considered some basic problems of the non-parametric estimation theory for discrete-time single-type branching processes with or without immigration. Some essential asymptotic results were presented in details. Of course, there is a variety of other research problems, directions and results in this area. Below we list some of them.

Assuming that offspring probabilities (2.1) depend on an unknown parameter, the parametric estimation theory is developed in many papers. Some of the results for the classical BGW process are given in the book of Guttorp (1991) (see also the references therein). Parametric estimation for BGWR processes is developed in Stoimenova and Yanev (2005).

Heyde and Lin (1992) and Heyde (1993, 1997) developed quasi-likelihood estimation methods with applications in branching processes.

Maximum-likelihood estimation for BGW process based on the sample (2.6) only is developed by Epps and Seneta (1992) in the case when individuals admit a bounded number of offspring. Another sampling scheme is considered by Maki and McDunnough (1989a, 1989b) where they select at random individuals from some generation and observe their family sizes.

The Bayes statistical methods are considered in the book of Guttorp (1991) (see also the references therein) where results for the classical BGW process are presented. Yanev and Tsokos (1999) obtained some new results in this direction. Johnson et al. (1979) developed a nonparametric Bayesian estimation in the case of age-dependent branching processes (note that these are non-Markov processes with continuous time).

Sequential estimation in BGWI processes is proposed by Sriram et al. (1991) and Qi and Reeves (2002).

An interesting observation scheme is considered in Jacob and Peccoud (1998) for estimating some parameters (especially the estimation of the initial size assuming that the age is known). Further development is given in Meester and Trapman (2006).

Jacob and Lalam (2004), Lalam and Jacob (2004) and Lalam et al. (2004) developed an estimation theory for size-dependent branching processes. Jacob et al. (2005) proposed statistical inferences for some classes of regenerative branching processes.

For the estimation problems and results in multitype BGW processes we refer to the books of Badalbaev and Mukhitdinov (1990) and Guttorp (1991), and to the excellent review of Dion (1993). Some new results in this direction are obtained in Chi (2004) and Maaouia and Touati (2005).

Statistical inference for bisexual branching processes is developed in the papers of Molina et al. (1998), González et al. (2001), Molina et al. (2005, 2007) and González et al. (2007). Note that the last three papers deal with the complex case of population-size dependent mating.

Estimation problems for processes with immigration are considered by Bercu (2001), Shete and Sriram (2003) and Ispány et al. (2005). Bootstrap estimation in BGWI is developed by Sriram (1994) and by Datta and Sriram (1995).

The statistical inference for continuous-time branching processes is a very important topic for investigations where there is a lot of open problems especially in the non-Markov cases. One can refer to Hoel and Crump (1974), Jagers (1975) and Athreya and Keiding (1977) for some basic results. Stimulated by modeling of cell proliferation with age-dependent branching processes, Yakovlev et al. (1998a,1998b, 2000), Zorin et al. (2000), Hyrien et al. (2005a, 2005b)), Hyrien (2007) and Hyrien and Zand (2007) developed an estimation theory with real data application to cell populations.

Robust estimation for BGW process is considered in Sriram and Vidyashankar (2000). Robust methods in BGWR processes are developed by Stoimenova et al. (2004a, 2004b), Stoimenova (2005) and Atanasov et al. (2007). A software system for simulation and estimation of branching processes with random migration is considered and discussed in Nitcheva and Yanev (2000). A new version and further development of this system in MATLAB environment is developed by Atanasov and Stoimenova (2007).

Estimators for controlled branching processes are proposed and investigated in the recent papers of González et al. (2004a, 2004b) and Sriram et al. (2007). Note that all popular models of discrete-time branching processes (in particular the BGW process with or without immigration) can be obtained as particular cases of the branching processes with random control functions.

Finally we would like to point out that a variety of branching models are studied from the probabilistic point of view. However the statistical results for these models are very limited (or even absent), which means that there is a lot of possibilities for further statistical investigations.

References

[1] Asmussen, S.; Hering, H. *Branching Processes*. Birkhauser: Boston, MA, 1983.

[2] Atanasov, D.; Stoimenova, V.; Yanev, N. Estimators in branching processes with immigration. *Pliska-Stud. Math. Bulg*. 2007, *18*, 19–40.

[3] Atanasov D.; Stoimenova V. Branching Processes Engine Rev. 2, 2007, http://www.fmi.uni-sofia.bg/fmi/statist/projects/bp.

[4] Athreya, K. B.; Keiding, N. Estimation theory for continuous-time branching processes. *Sankhya* (Ser. A) 1977, *39*, 2:101–123.

[5] Athreya, K. B.; Ney, P. E. *Branching Processes*. Springer: Berlin, 1972.

[6] Badalbaev, I. S.; Mukhitdinov, A. *Statistical Problems in Multitype Branching Processes*. Fan: Tashkent, 1990 (in Russian).

[7] Basawa, I. V.; Scott, D. J. *Asymptotic Optimal Inference for Non-Ergodic Models.* Springer: New York, 1983.

[8] Bercu, B. Weighted estimation and tracking for branching processes with immigration. *IEEE Trans. on Automat. Control* 2001, *16*, 43–50.

[9] Bühler, W. J. Ein zentraler Grenzwertsatz fur Verzweigungsprozesse. *Z. Wahrscheinlichkeitstheor. Verw. Geb.* 1969, *11*, 139–141.

[10] Chauvin, B. Sur la proprieté de branchment. *Ann. Inst. Henri Poincaré* 1986, *22*, 233–236.

[11] Chi, Z. Limit laws of estimators for critical multi-type Galton-Watson process. *Ann. Appl. Prob.* 2004, *14*, 1992–2015.

[12] Crump, K. S.; Howe, R. B. Nonparametric estimation of the age of a Galton-Watson branching processes. *Biometrika* 1972, *59*, 533–538.

[13] Datta, S.; Sriram, T. N. A modified bootstrap for branching processes with immigration. *Stoch. Proc. Appl.* 1995, *56*, 275–295.

[14] Dion, J.-P. Estimation of the mean and the initial probabilities of a branching process. *J. Appl. Prob.* 1974, *11*, 687–694.

[15] Dion, J.-P. Estimation of the variance of a branching process. *Ann. Statist.* 1975, *3*, 1183–1187.

[16] Dion, J.-P. Statistical inference for discrete time branching processes. In Obretenov, A.; Stefanov, V. T., (Eds.), *Lecture Notes of 7th Int. Summer School on Prob. Theory and Math. Statistics.* Science Culture Technology Publishing, Singapore, 1993, 60–121.

[17] Dion, J.-P.; Keiding, N. Statistical inference in branching processes. In: Joffe, A.; Ney, P., (Eds.) *Branching processes.* Dekker: New York, 1978, 105–140.

[18] Dion, J.-P.; Yanev, N. M. A new transfer limit theorem. *C. R. Acad. Bulg. Sci.* 1991, *44*, 1:19–21.

[19] Dion, J.-P.; Yanev, N. M. Limiting distributions of Galton-Watson branching processes with a random number of ancestors. *C. R. Acad. Bulg. Sci.* 1991, *44*, 3:23–26.

[20] Dion, J.-P.; Yanev, N. M. Estimation theory for branching processes with or without immigration. *C. R. Acad. Bulg. Sci.* 1991, *44*, 4:19–22.

[21] Dion, J.-P.; Yanev, N. M. Estimation theory for the variance in branching processes with a random number of ancestors. *C. R. Acad. Bulg. Sci.* 1992, *45*, 11:27–30.

[22] Dion, J.-P.; Yanev, N. M. Statistical inference for branching processes with censored observations. *C. R. Acad. Bulg. Sci.* 1992, *45*, 12:21–24.

[23] Dion, J.-P.; Yanev, N. M. Statistical inference for branching processes with an increasing random number of ancestors. *J. Statist. Planning & Inference* 1994, *39*, 329–352.

[24] Dion, J.-P.; Yanev, N. M. CLT for martingales in BGWR branching processes with some statistical applications. *Math. Methods of Statistics* 1995, *4*, 344–358.

[25] Dion, J.-P.; Yanev, N. M. Limit theorems and estimation theory for branching processes with an increasing random number of ancestors.*J. Appl. Prob.* 1997, *34*, 309–327.

[26] Duby, C.; Rouault, A. Estimation simultanée de l'espérance et de la variance pour un processus de Galton-Watson. *C. R. Acad. Sci.* 1980, AB*290*, 7:A339–A341.

[27] Duby, C.; Rouault, A. Estimation non-paramétrique de l'espérance et de la variance de la loi de reproduction d'un processus de ramification.*Ann. Inst. Henri Poincaré* 1982, *18*B, 149–163.

[28] Epps, T. W.; Seneta, E. Maximum-likelihood estimation for Galton-Watson processes. *Comm. Statist. Theory & Methods* 1992, *21*, 733–748.

[29] Feigin, P. D. A note on maximum likelihood estimation for simple branching processes. *Austr. J. Statist.* 1977, *19*, 152–154.

[30] González, M.; Molina, M.; Mota, M. Estimation of the offspring distribution and the mean vector for a bisexual Galton-Watson process.*Comm. Statist. Theory & Methods* 2001, *30*, 497–516.

[31] González, M.; Martinez, R.; del Puerto, I. Non-parametric estimation of the offspring distribution and the mean for a controlled branching process.*Test* 2004, *13*, 465–479.

[32] González, M., Martinez; R. del Puerto, I. Estimation of the variance for a controlled branching process. *Test* 2005, *14*, 199–213.

[33] González, M.; Mota, M.; Ramos, A. Estimation in a class of bisexual process with population-size dependent mating.*Austral.& N. Zeal. J. Statist.* 2007, *49*, 37–50.

[34] Guttorp, P. *Statistical inference for Branching Processes.* Wiley: New York, 1991.

[35] Haccou, P.; Jagers, P.; Vatutin, V. A. *Branching Processes: Variation, Growth and Extinction of Populations.* Cambridge University Press: Cambridge, 2005.

[36] Harris, T. E. Branching Processes. *Ann. Math. Statist.* 1948, *19*, 474–494.

[37] Harris, T. E. *Branching Processes*. Springer: New York, 1963.

[38] Heathcote, C. R. A branching process allowing immigration. *J. Royal Statist. Soc.* 1965, *1*, 138–143; *Corrections* 1966, *1*, 213–217.

[39] Heyde, C. C. Extension of a result of Seneta for the supercritical Galton-Watson process. *Ann. Math. Statist.* 1970, *41*, 739–742.

[40] Heyde, C. C. On estimating the variance of the offspring distribution in a simple branching process. *Adv. Apl. Prob*. 1974, *3*, 421–433.

[41] Heyde, C. C. Remarks on efficiency in estimation for branching processes. *Biometrika* 1975, *62*, 49–55.

[42] Heyde, C. C. Quasi-lkelihood and general theory inference for stochastic processes. In Obretenov, A.; Stefanov, V.T., (Eds.) *Lecture Notes of 7th Int. Summer School on Prob. Theory and Math. Statistics*. Science Culture Thechnology Publishing, Singapore, 1993, 122–152.

[43] Heyde, C. C. Quasi-Likelihood and Its Applications. *A General Approach to Optimal Parameter Estimation*. Springer: New York, 1997.

[44] Heyde, C. C.; Lin, Y. X. On quasi-likelihood methods and estimation for branching processes and heteroscedastic regression models. *Austral. J. Statist*. 1992, *34*, 199–206.

[45] Heyde, C. C.; Seneta, E. Estimation theory for growth and immigration rates in a multiplicative process. *J. Appl. Prob*. 1972, *9*, 235–256.

[46] Heyde, C. C.; Seneta, E. Notes on "Estimation theory for growth and·immigration rates in a multiplicative process". *J. Appl. Prob*. 1974, *11*, 572–577.

[47] Hoel, D. G.; Crump, K. S. Estimating the generation-time distribution of an age-dependent branching process. *Biometrics* 1974, *30*, 125–135.

[48] Hyrien, O. Pseudo-likelihood estimation for discretely observed multitype Bellman-Harris branching processes. *J. Statist. Planning & Inference* 2007, *137*, 1375–1388.

[49] Hyrien, O.; Mayer-Proschel, M.; Noble, M.; Yakovlev, A. Y. A stochastic model to analyze clonal data on multi-type cell populations. *Biometrics* 2005, *61*, 199–207.

[50] Hyrien, O.; Mayer-Proschel, M.; Noble, M.; Yakovlev, A. Y. Estimating the life-span of oligodendrocytes from clonal data on their development in cell culture. *Math. Biosci*. 2005, *193*, 255–274.

[51] Hyrien O.; Zand M. S. A mixture model with dependent observations for the analysis of CFSE-labeling experiments. *JASA* 2007 (in press).

[52] Iosifescu, M.; Grigorescu, S.; Oprisan, G.; Popescu, G. *Elemente de Modelare Stochastica*. Editura Tehnica: Bucaresti, 1984 (in Romanian).

[53] Ispány, M.; Pap, G.; van Zuijlen, M. C. A. Fluctuation limit of branching processes with immigration and estimation of the means. *Adv. Appl. Prob*. 2005, *37*, 523–538.

[54] Jacob, C. Peccoud, J. Estimation of the parameters of a branching process from migrating binomial observations. *Adv. Appl. Prob*., 1998, *30*, 948–967.

[55] Jacob, C.; Lalam, N. Estimation of the offspring mean in a general single-type size-dependent branching processes. *Pliska-Stud. Math. Bulg*. 2004, *16*, 65–88.

[56] Jacob, C.; Lalam, N.; Yanev, N. M. Statistical inference for processes depending on environments and applications in regenerative processes. *Pliska-Stud. Math. Bulg.* 2005, *17*, 109–136.

[57] Jagers, P. *Branching Processes with Biological Applications.* Wiley: London, 1971.

[58] Johnson, R. A.; Susarla, D.; van Rysian. Bayesian non-parametric estimation for age-dependent branching processes. *Stoch. Proc. Appl.* 1979, *9*, 307–318.

[59] Keiding, N.; Lauritzen, S. Marginal maximum likelihood estimates and estimation of the offspring mean in branching process. Scand.*J. Statist.* 1978, *5*, 106–110.

[60] Kimmel, M.; Axelrod, D. E. *Branching Processes in Biology.* Springer: New York, 2002.

[61] Kolmogorov, A. N. On the solution of a biological problem. *Proc. of Tomsk Univ.* 1938, *2*, 7–12 (in Russian).

[62] Lalam, N.; Jacob, C. Estimation of the offspring mean in a supercritical or nearcritical branching processes. *Adv. Appl. Prob.* 2004, *36*, 582–601.

[63] Lalam, N.; Jacob, C.; Jagers, P. Modeling of the PCR amplification process by a size-dependent branching process and estimation of the efficiency. *Adv. Appl. Prob.* 2004, *36*, 602–615.

[64] Lotka, A. J. Theorie analitique des associations biologiques. *Act. Sci. Indust.* 1939, *2*, 123–136.

[65] Maaouia, F.; Touati, A. Identification of multitype branching processes. *Ann. Statist.* 2005, *33*, 2655–2694.

[66] Maki, E.; McDunnough, P. Sampling a branching tree. *Stoch. Proc. Appl.* 1989, *31*, 383–305.

[67] Maki, E.; McDunnough, P. What can and what can't be estimated in branching and related processes? *Stoch. Proc. Appl.* 1989, *31*, 307–314.

[68] Meester, R.; Trapman, P. Estimation in branching processes with restricted observations. *Adv. Appl. Prob.* 2006, *38*, 1098–1115.

[69] Melnikov, A. V.; Novikov, A. A. On sequel inference with guaranteed accuracy for semi-martingales. *Theory Probab. Appl.* 1988, *33*, 480–494.

[70] Mode, C. J. *Multitype Branching Processes: Theory and Applications.* Elsevier: New York, 1971.

[71] Molina, M.; Gonzales, M.; Mota, M. Bayesian inference for bisexual Galton-Watson process. *Comm. Statist. Theory & Methods* 1998, *27*, 1055–1070.

[72] Molina, M.; Mota, M.; Ramos, A. Non-parametric estimation in a class of bisexual process with population-size dependent mating. *Pliska-Stud. Math. Bulg.* 2005, *17*, 155–180.

[73] Molina, M.; Mota, M.; Ramos, A. Bayesian estimation in a class of bisexual process with population-size dependent mating. *Test* 2007 (to appear).

[74] Nanthi, K. Statistical Estimation for Stochastic Processes. *Queen's Papers in Pure and Aplied Mathematics*, 1983, *62*.

[75] Nagaev, A. V. On estimating the expected number of direct descendants of a particle in a branching process. *Theory Probab. Appl.* 1967, *12*, 314–320.

[76] Nitcheva, D.; Yanev, N. M. A system for simulation and estimation of branching processes. *Pliska-Stud. Math. Bulg.* 2000, *13*, 259–271.

[77] Pakes, A. G. Non-parametric estimation in the Galton-Watson process. *Math. Biosciences* 1975, *26*, 1–18.

[78] Qi, Y.; Reeves, J. On sequential estimation for branching processes with immigration. *Stoch. Proc. Appl.* 2002, *100*, 41–51.

[79] Quine, M. P. Asymptotic results for estimators in a subcritical branching process with immigration. *Ann. Probab.* 1976, *4*, 319–325; *correction*, 1977, *5*, 318.

[80] Sankaranarayanan, G. *Branching Processes and its Estimation Theory.* Wiley: New York, 1989.

[81] Sevastyanov, B. A. *Branching Processes.* Nauka: Moscow, 1971 (in Russian).

[82] Sevastyanov, B. A. Limit theorems for special type branching processes. *Theory Probab. Appl.* 1957, *2*, 339–348.

[83] Scott, D. J. A central limit theorem for martingales and application to branching processes. *Stoch. Proc. Appl.* 1978, *6*, 241–152.

[84] Shete, S.; Sriram, T. N. A note on estimation in multitype supercritical branching processes with immigration. *Sankhya* 2003, *65*, 107–121.

[85] Sriram, T. N. Invalidity of bootstrap for critical branching processes with immigration. *Ann. Statist.* 1994, *22*, 1013–1023.

[86] Sriram, T. N.; Basawa, I. V.; Huggins, R. M. Sequential estimation for branching processes with immigration. *Ann. Statist.* 1991, 2232–2243.

[87] Sriram, T. N.; Vidyashankar, A. N. Minimum Hellinger distance estimation for supercritical Galton-Watson processes. *Statist. Probab. Letters* 2000, *50*, 331–342.

[88] Sriram, T. N.; Bhattacharya, A.; Gonzalez, M.; Martinez, R.; del Puerto, I. Estimation of the offspring mean in a controlled branching process with a random control function. *Stoch. Proc. Appl.* 2007 (to appear).

[89] Stigler, S. M. Estimating of the age of a Galton-Watson branching processes. *Biometrika* 1970, *57*, 505–512.

[90] Stoimenova, V. Robust parametric estimation of branching processes with random number of ancestors. *Serdica Math. J.* 2005, *31*, 243–262.

[91] Stoimenova, V.; Atanasov, D.; Yanev, N.M. Robust estimation and simulation of branching processes. *C. R. Acad. Bulg. Sci.* 2004, *55*, 19–22.

[92] Stoimenova, V., Atanasov, D. Yanev, N.M. Simulation and robust modifications of estates in branching processes. *Pliska-Stud. Math. Bulg.*, 2004, *16*, 259–271.

[93] Stoimenova, V.; Yanev, N. M. Parametric estimation in branching processes with an increasing random number of ancestors. *Pliska-Stud. Math. Bulg.* 2005, *17*, 295–312.

[94] Vatutin, V. A.; Zubkov, A. M. Branching Processes, Part 1. *Itogi Nauki i Tehniki* 1985, *23*, 3–67 (in Russian).

[95] Wei, C. Z.; Winnicki, J. Some asymptotic results for branching processes with immigration. *Stoch. Appl.* 1989, *31*, 26–282.

[96] Wei, C. Z.; Winnicki, J. Estimation of the means in the branching processes with immigration. *Ann. Statist.* 1990, *18*, 1757–1773.

[97] Winnicki, J. Estimation theory for the branching processes with immigration. *Contemporary Mathematics* 1988, *80*, 301–322.

[98] Winnicki, J. Estimation of the variances in the branching processes with immigration. *Prob. Theory & Rel. Fields* 1991, *88*, 77–106.

[99] Yaglom, A. M. Some limit theorems in the theory of branching random processes. *Proc. Acad. Sci. USSR* 1947, *56*, 795–798 (in Russian).

[100] Yakovlev, A. Y.; Yanev, N. M. Transient Processes in Cell Proliferation Kinetics. *Lecture Notes in Biomathematics,* 82. Springer: Berlin, 1989.

[101] Yakovlev, A. Y.; Boucher, K.; Mayer-Proschel, M.; Noble, M. Quantitative insight into proliferation and differentiation of oligodentrocyte-type 2 astrocyte progenitor cells in vitro. *Proc. Natl. Acad. Sci. USA* 1998, *95*, 4:14164–14167.

[102] Yakovlev, A. Y.; Mayer-Proschel, M.; Noble, M. A stochastic model of brain cell differentiation in tissue culture. *J. Math. Biol.* 1998, *37*, 49–60.

[103] Yakovlev, A. Y.; Mayer-Proschel, M.; Noble, M. Stochastic formulations of a clock model for temporally regulated generation of oligodendrocytes in vitro. *Math. Compute. Model.* 2000, *32*, 125–137.

[104] Yanev, G. P.; Tsokos C. P. Decision-theoretic estimation of the offspring mean in mortal branching processes. *Stoch. Models* 1999, *15*, 889–902.

[105] Yanev, N. M. On the statistics of branching processes. *Theory Probab. Appl.* 1975, *3*, 623–633.

[106] Yanev, N. M. Limit theorems for estimators of individual characteristics in a Galton-Watson process. *Serdica Bulg. Math. Publ.* 1986, *3*, 143–153.

[107] Yanev, N. M.; Tchoukova-Dantcheva, S. On the statistics of branching processes with immigration. *C. R. Acad. Bulg. Sci.* 1980, *33*, 469–472.

[108] Yanev, N. M.; Tchoukova-Dantcheva, S. Limit theorems for estimators of variances in a branching process with immigration. *Serdica Bulg. Math. Publ.* 1986, *12*, 3:134–142 (in Russian).

[109] Zorin, A. A.; Yakovlev, A. Y.; Mayer-Proschel, M.; Noble, M. Estimation problems associated with stochastic modeling of proliferation and differentiation of O-2A progenitor cells in vitro. *Math. Biosci.* 2000, *167*, 109–121.

INDEX

G

I

K

L

M

N

O

P

R

S

T

U

V

W